CROCK-POT

EXPRESS ® DUMP MEALS

COOKBOOK

Delicious recipies that are simple, quick and easy
to make. Prep Time 10 Minutes or Less

Mary E. WAtson

Copyright

TABLE OF CONTENTS

INTRODUCTION

Crock-Pot Express® Dump Meals Cookbook will provide you with delicious cooking solutions every single day. If you don't have a lot of time to spend in the kitchen, worried about cooking a bad meal, or are not well-versed with cooking, these recipes are for you!

Extremely Short Prep Times- Mae sure that you note the prep time mentioned at the head of every single recipe.

Novice Book- Every single step that you need to take to prepare each of the meal in this book is very easily explained to make sure that you are able to understand them even if you do not have a lot of cooking experience.

Family Recipes- The recipes have not come from famous chefs but from home cooks. The recipes are the favorites of these home cooks and are cherished by families all over the world. So get your Crock-Pot Express Multi-Cooker® and prepare yourself to cook some simple, flexible and absolutely amazing recipes.

Only use the ingredients that you already have- You can choose a recipe that suits your taste buds or of those whom you want to please. You can even go ahead and cook recipes that only need the ingredients that you already have. There are many different types of recipes in this book and many of them only differ from others by an ingredient or probably two.

Grandma TIPS- You will also find many useful TIPS throughout this cookbook. These are the tips that one learns after using a Crock-Pot Express Multi-Cooker® for a long period of time or by cooking on a regular basis.

Less Cook and More Talk- Crock-Pot Express Multi-Cooker® have proved time and again that they are a perfect friend of those who are not at home throughout the day or people who don't want to spend long, rigorous hours in the kitchen, but still want to offer delicious meals to the family members.

Time Management- With a Crock-Pot Express Multi-Cooker®, you can make a meal one evening, store the lift-out part of the cooker in the refrigerator overnight, and then place it again into the electronic component of the cooker in the morning. You can even simply carry all of it as a carry-in meal or a buffet, while the quality of the contents still remains intact.

Why should you choose **Crock-Pot Express® Dump Meals Cookbook**?

- Preparations are **super-easy** and the required ingredients are easily available at reasonable costs

- Many of the recipes only need a **few ingredients** and they too are readily available
- Everyone can **enjoy** the meal, including the cook as he/she won't be spending a lot of time or effort for cooking
- Fully loaded with recipes
- Many of the **ingredient**s are generally **available at home**
- Can **easily choose one** that suits you best
- **Delicious flavors** with minimum fuss
- No matter how crazy or tiring your day was, you can **relax and enjoy delicious** meals with the ones you love with minimum efforts

! Dump…. and Enjoy !!!!!

TIPS FOR NEWBIES

1. When you start, the pressure valve on top must be set to Locked position, not open. This means that the pot is sealed shut and steam cannot come out. Therefore, the pressure can build up.
2. Then make sure the lid is closed tightly. Before you close it, make sure that the white silicone ring on the inside of the lid is in place all the way around.
3. Choose which button you want to push, depending on your recipe and push it.
4. Next, the number of minutes that your thing will cook will show up and it will start counting down.
5. When it gets to "0", you can leave it alone and let it "NPR" or "Naturally Release the Pressure", for as long as you like or as long as your recipe specifies. During this time, the numbers will go back up and these numbers tell you how many minutes it has been in the "NPR" state. The metal pin will drop down into the lower position when it is safe to open it.
6. If you want to let the pressure out quickly, right when it's done or any time later on, turn the pressure valve to "Open Position" and get your hands out of the way. Steam will come shooting out of the float valve. When it stops shooting out and the metal pin drops down to the lower position, it is safe to open.
7. The "timer" button is NOT a timer button! It is really a "delay start" button and is only used for setting the amount of delay you want before the cooking starts.
8. If you have a recipe for unfrozen meat, and your meat is frozen, use the same cooking time and amount of liquid indicated in your recipe. The difference is that the time for the Crock-Pot Multi-Cooker ® to come to pressure will increase.

40 IMPORTANT CROCK-POT MULTI-COOKER ® TIPS

1. It is a good idea to always keep a second/back-up silicone ring/gasket on hand. They usually last at least a couple of years, but things happen, and without a gasket your Crock-Pot Multi-Cooker® is out of commission.
2. The ring (as well as the float valve assembly) can be steam cleaned by adding a couple of cups of water to the liner along with a cut up lemon, and steam set on "Steam" for 3 minutes. You can then use this lemon water to clean other things as well.
3. Did you know the "Control Panel" (keypad on the front of your Crock-Pot Multi-Cooker®) comes with a protective film? It is nearly undetectable, and pretty durable, but over time/use it will develop cracks/bubbles. This film is meant to be removed, and some people mistake this cracking for a problem with their keypad. Peel this film off and reveal a fresh, clean, durable surface!
4. Always dry the bottom and sides of the liner before putting it in the base. General safety precaution.
5. Check that your "condensation collector" cup is attached to the back, and empty/wash it regularly.
6. Don't add too much liquid! This causes the pot to take MUCH longer to get to pressure and much longer to come down - and it drowns food (not for rice or grains)
7. Don't cook food longer than necessary, use a reliable pressure-cooking chart, and keep your own records too!
8. Use the Time Adjustment Button, If you can use only 1 button on the Crock-Pot Express Multi-Cooker ®, it'll be the Time Adjustment Button. Most of the recipes are developed using the Time Adjustment Button, because it gives us control to cook precisely and accurately.
9. Quick Release (QR) vs. Natural Release (NR) most common ways to release pressure in Crock-Pot Multi-Cooker ®: Quick Release & Natural Release.
10. Thick Liquid Generates Less Steam, If the pressure-cooking liquid is too thick, Crock-Pot Multi-Cooker ® may not be able to reach the desired pressure to start the pressure-cooking cycle. Consider thinning out the liquid by adding chicken stock or water.
11. Safest Way to Open the Lid, The safest way to open the lid is to slowly tilt it away from you. The lid acts as a guide to direct the steam away from you. Accidents do happen in kitchen. In extremely rare situation (user error), the lid also acts as a shield to protect you.
12. How to Double the Recipes; for most dishes, you do not need to increase the cooking time when doubling a recipe. Please keep this Important Safeguard in mind when doubling a recipe: Do Not Overfill – Overfilling may clog the Pressure Release Valve and develop excess pressure. All Pressure Cooking Programs: Do not fill the unit over 2/3 full. For Food that Expands During Cooking (i.e. grains, beans, and dried vegetables): Do not fill the unit over 1/2 full.
13. Pot-in-Pot (PIP): What & How? Pot-in-Pot = 2 in 1 (2 dishes in 1 pot). The Pot-in-Pot method allows you to cook 2 dishes separately in Crock-Pot Multi-Cooker ® at the

same time. Example: You can cook chicken thighs at the bottom of the pot, and a bowl of rice on top with a steamer rack lifting it up (separating it from the chicken).

14. Always Check the Sealing Ring. The silicone sealing ring deforms over time. Develop the habit to check every time if the sealing ring is properly seated before pressure cooking. Crock-Pot Multi-Cooker® recommends replacing the Silicone Sealing Ring every 18 – 24 months or when you notice deformations.

15. Easy Recipes to Get You Started. There are more things you can make with your new kitchen toy – Crock-Pot Multi-Cooker® – than you've imagined!

16. Use stop-and-go cooking for perfect results. When making a recipe that contains ingredients that cook at different times, begin by partially cooking slow-to-cook foods, such as meat, first. Then use a quick-release method to stop the pressure cooker. Next, add the faster-cooking ingredients — such as green beans or peas — to the meat. Bring the pot back up to pressure again and finish everything up together at the same time.

17. Bear in mind that high altitude means longer cooking times. You may have to increase the cooking times if you live at an elevation of 3,000 feet above sea level or higher. A good general rule is to increase the cooking time by 5 percent for every 1,000 feet you are above the first 2,000 feet above sea level.

18. Release that pressure. When the food is done cooking under pressure, use the appropriate pressure-release method mentioned above, according to the recipe you're making.

19. Make a perfect Popcorn in the Electric Pressure Cooker, use the "Meat/Stew" function and covers the pot with the glass lid designed for slow cooking. Heat the oil, if you are using coconut oil; allow all of the solid oil to melt. Put 3 or 4 popcorn kernels into the oil. When the kernels pop, add the rest of the 1/3 cup of popcorn kernels in an even layer. Cover, **turn off** and count 30 seconds. (Count out loud; it's fun to do with kids.)

20. You can cook food directly from its frozen state! You can skip the thawing step and still have your freezer meals on the table in record time. But there is one catch to cooking freezer meals in the Crock-Pot Multi-Cooker®. Because the appliance is round you cannot freeze your meals in the bag flat. Place your freezer bag into bowls, pots, pans, etc. that will fit inside your Crock-Pot Multi-Cooker ® and freeze them in that mold. Allow your Crock-Pot Multi-Cooker® extra time when cooking from frozen.

21. Do the familiarization test ('water test') found in this book and in the user manual – it really will help you to feel comfortable with the controls and the way your Crock-Pot Multi-Cooker® functions and takes care of itself.

22. Check that the sealing ring float valve and steam release knob are in position, clean and operating correctly before each use!

23. Add liquid when pressure-cooking. Your Crock-Pot Multi-Cooker® requires steam to create pressure. It will need a minimum of 1 measuring cup of water, stock or other liquid in order to make the vapor to cause pressure.

24. Thicken a sauce at the end of the cooking process. The addition of flour, stock cubes or other 'thickeners' at the beginning of the cooking process can cause a starch layer to form on the base of the inner pot and for the Crock-Pot Multi-Cooker® to display an overheat message.

25. If using a stove-top recipe for your Crock-Pot Multi-Cooker® that has been written for a 15psi 'stove top' pressure cooker add 5% to the cooking time.
26. If you are using your Crock-Pot Multi-Cooker® at altitude you need to add 5% to the recipe cooking time for every 1000 feet above sea level.
27. If your dish is overcooked you can't do anything about it. If not certain about a timing, reduce the time; you can always cook it for a further period of time if needed. Take Notes Learn From Your Successes And From Your Failures!
28. Sauté first for increased flavour. Onions and vegetables can be caramelized in the inner pot, meat can be browned all before you add additional ingredients and commence the cooking program.
29. After using the sauté option and adding the other ingredients for a pressure-cooking program, give everything a good stir and make sure any caramelized ingredients are scraped of the inner pot base.
30. Evaporation is minimal when using the Crock-Pot Multi-Cooker® so reduce the liquid content accordingly and don't forget all liquid counts – so if using something like a tin of chopped tomatoes reduce the liquid content accordingly.
31. Write down the process if it goes well so you can repeat it. Part of the fun of owning your Crock-Pot Multi-Cooker® is learning a new way to cook, if you have success write it down so you can repeat it.
32. Don't forget the keep warm period. The keep warm part of a cooking program is at 68c / 154f and as such food will continue to be cooked during the 10-15 minutes a natural pressure release may occur. Take that time into account when planning your meal.
33. If you forget to soak your dry beans use the quick soak technique. Rinse then add to the Crock-Pot Multi-Cooker®, cover with water and pressure cook for 2 minutes. Natural pressure release, rinse and then use in a recipe as if soaked.
34. Be careful when moving the cooker and placing the lid down in your kitchen, hot hobs will melt the base unit and it is very easy to do.
35. Before placing the inner pot inside the Crock-Pot Multi-Cooker®, check that the base is clean and clear of all debris and that it is dry especially under the rim.
36. If you need to change temperatures when using the slow cook or the sauté function use the 'stop' button and restart the process at the correct temperature. Any cooking program can be stopped via the 'stop' button so don't worry.
37. The Crock-Pot Multi-Cooker® sealing ring can be safely cleaned in the dishwasher. To remove stubborn odors from the sealing ring, cut up half a lemon and place it in the cooking pot (with the sealing ring in normal position) with 2 cups of water. Put on the lid, close the steam release valve and select "steam" for 3 minutes.
38. You can use the pre-set temperatures to be even more adventurous. The Crock-Pot Multi-Cooker® keep warm; Slow cook programs can be used for a variety of tasks.
39. Be prepared to buy 2 – or at the very least a second inner pot – you will very quickly find that a dessert can be cooking while a main meal is served, or vegetables can quickly cook while your main is 'resting' – a second inner pot can be invaluable to make the most of your Crock-Pot Multi-Cooker®.

Apples And Cinnamon Oatmeal

PREP: 5 MINUTES • PRESSURE: 12 MINUTES • TOTAL: 17 MINUTES • PRESSURE LEVEL: HIGH • RELEASE: NATURAL

SERVES: 4

Ingredients

3 cups water
2 tablespoons packed brown sugar
½ teaspoon ground cinnamon
¼ teaspoon kosher salt
¾ cup steel-cut oats
1 small apple, peeled, cored, and diced
1 teaspoon unsalted butter
1-tablespoon heavy (whipping) cream

Directions

1. **Preparing the Ingredients.** In the Crock-Pot Multi-Cooker®, stir together the water, brown sugar, cinnamon, and kosher salt, dissolving the salt and sugar. Pour in the oats, add the apple, and stir again.
2. **High pressure for 12 minutes.** Lock the lid in place, cook for 12 minutes. To get 12 minutes cook time, press the "Rice/Risotto" button. When the time is up turn the Crock-Pot Multi-Cooker off. ("Keep warm" setting, turn off).
3. **Pressure Release**. Use the natural release method. Unlock and open the Crock-Pot Multi-Cooker ®.
4. **Finish the dish**. Stir the oats, and taste; if you like them softer, place the lid on the cooker, but *don't lock* it. Let the oats sit for 5 to 10 minutes more. When they are ready to serve, stir in the butter and heavy cream.
5. Serve and Enjoy!

PER SERVING: CALORIES: 181; FAT: 4G; SODIUM: 157MG; CARBOHYDRATES: 31G; FIBER: 4G; PROTEIN: 5G

Banana Oatmeal

PREP: 5 MINUTES • PRESSURE: 18 MINUTES • TOTAL: 24 MINUTES • PRESSURE LEVEL: HIGH • RELEASE: NATURAL

SERVES: 4

Ingredients

½ cup steel-cut oats

½ cup packed light brown sugar

2 ripe bananas, chopped

2 teaspoons vanilla extract

½ teaspoon ground cinnamon

¼ teaspoon salt

¼ cup heavy cream

Directions

1. **Preparing the Ingredients.** Mix the oats, brown sugar, bananas, vanilla, cinnamon, and salt with 2¼ cups water in the Crock-Pot Multi-Cooker® until the brown sugar dissolves.
2. **High pressure for 18 minutes**. Lock the lid onto the pot and cook at high pressure for 18 minutes. To get 18 minutes cook time, press the "Poultry" button and use the TIME ADJUSTMENT button to adjust the cook time to18 minutes.
3. **Pressure Release**. Turn off the Crock-Pot Multi-Cooker® or unplug it so it doesn't flip to its keep-warm setting. Allow the pot's pressure to come to normal naturally, 10 to 12 minutes.
   If the pot's pressure hasn't returned to normal within 12 minutes, use the quick-release method to bring it back to normal.
4. **Finish the dish** Unlock and open the cooker. Stir in the cream and set aside for 1 minute to warm before serving.

Egg and Cheese Breakfast

PREP: 5 MINUTES • PRESSURE: 4 MINUTES • TOTAL: 9 MINUTES • PRESSURE LEVEL: HIGH • RELEASE: QUICK
SERVES: 2

Ingredients

1 teaspoon-unsalted butter, at room temperature, divided
2 large eggs
¼ teaspoon kosher salt, divided
Freshly ground black pepper
2 tablespoons grated aged Cheddar or Parmesan cheese, divided
1-cup water, for steaming
2 English muffins

Directions

1. **Preparing the Ingredients.** Using ½ teaspoon of butter each, coat the insides of 2 heatproof custard cups or small ramekins. Crack 1 egg into each cup, and carefully pierce the yolks in several places to make sure the yolk cooks through evenly. Sprinkle each with ⅛ teaspoon of kosher salt, some pepper, and 1 tablespoon of Cheddar cheese, covering the eggs. Cover the cups with aluminum foil, crimping it around the sides.
   Add water and insert the steamer basket or trivet. Place the cups on the insert.
2. **High pressure for 4 minutes**. Lock the lid in place, and bring the pot to high pressure for 4 minutes. To get 4-minutes cook time, press the Steam button and use the COOK TIME SELECTOR button to adjust the cook time to 4 minutes.
3. **Pressure Release** After the timer reaches 0, the cooker will automatically enter Keep warm mode. Press the Stop button and carefully release the pressure.
4. **Finish the dish** Toast the English muffins while the eggs cook.
   Unlock but *don't remove* the lid for another 30 seconds; this helps ensure that the whites are fully cooked. Using tongs, remove the cups from the cooker and peel off the foil.
   Using a small offset spatula or knife, loosen the eggs, then tip each one out onto the bottom half of one of the English muffins.
5. Top with the other half, and enjoy.

PER SERVING: CALORIES: 241; FAT: 9G; SODIUM: 682MG; CARBOHYDRATES: 26G; FIBER: 2G; PROTEIN: 14G

Bulgur, Oat, And Walnut Porridge

PREP: 5 MINUTES • PRESSURE: 25 MINUTES • TOTAL: 29 MINUTES • PRESSURE LEVEL: HIGH • RELEASE: QUICK

SERVES: 6

Ingredients

½ cup steel-cut oats
½ cup bulgur
½ cup chopped walnuts
½ cup maple syrup
½ teaspoon ground cinnamon
½ teaspoon salt

Directions

1. **Preparing the Ingredients.** Mix everything with 4 cups water in the Crock-Pot Multi-Cooker®.
2. **High pressure for 25 minutes.** Close the lid and the pressure valve and then cook for 25 minutes. To get 25-minutes cook time, press Beans/Chili button and use the TIME ADJUSTMENT button to adjust the cook time to 25 minutes.
3. **Pressure Release.** Use the quick-release method to bring the pot's pressure back to normal.
4. **Finish the dish**. Unlock and remove the lid. Turn the electric cooker to its browning function. Bring to a simmer, stirring often. Cook, stirring constantly, until slightly thickened, about 2 minutes.
5. Serve and Enjoy!

French Toast Bread Pudding

PREP: 5 MINUTES • PRESSURE: 15 MINUTES • TOTAL: 20 MINUTES • PRESSURE LEVEL: HIGH • RELEASE: QUICK
SERVES: 6

Ingredients
2 large eggs, at room temperature
1 cup whole or low-fat milk
¼ cup sugar
¼ cup orange marmalade
2 teaspoons vanilla extract
½ teaspoon ground cinnamon
5 cups of 1-inch bread cubes (about 7 ounces)
¼ cup raisins

Directions

1. **Preparing the Ingredients.** Lightly butter a 2-quart, high-sided, round baking or soufflé dish; set aside. Place the pressure cooker rack inside the Crock-Pot Multi-Cooker®; pour in 2 cups water.
   Whisk the eggs, milk, sugar, marmalade, vanilla, and cinnamon in a big bowl until smooth, with no bits of egg visible. Add the bread cubes and raisins; toss well to soak up the liquids. Pour the entire mixture into the prepared baking dish; cover and seal the dish with aluminum foil. Make a foil sling, set the filled baking dish on it, and lower the baking dish in the sling onto the rack. Fold the ends of the sling so they'll fit inside the cooker.
2. **High pressure for 15 minutes.** Lock the lid onto the cooker, bring the cooker to high pressure by pressing the Poultry button and cook for 15 minutes.
3. **Pressure Release.** Use the quick-release method to bring the pot's pressure back to normal.
4. **Finish the dish**. Unlock the lid and open the cooker. Use the foil sling to transfer the hot baking dish to a wire rack. Uncover and cool for 5 minutes before dishing it up by the big spoonful.

Soft, Medium, And Hard-Boiled Eggs

PREP: 5 MINUTES • PRESSURE: 3 MINUTES • TOTAL: 8 MINUTES • PRESSURE LEVEL: HIGH • RELEASE: VARIOUS
SERVES: 1-12

Ingredients
1–12 cold, large eggs (straight from the refrigerator)

Directions

Preparing the Ingredients. Set a large metal vegetable steamer in the Crock-Pot Multi-Cooker®; Add about 2 inches of water to the cooker—not so much that it comes through the holes of the steamer. Set one or more eggs in the steamer.

For soft-boiled eggs—Lock the lid onto the pot.

High pressure for 1 1/2 minutes. Bring the cooker to high pressure by pressing the STEAM button. Allow to cook for 1 1/2 minute and press START/STOP.

Pressure Release. Use the quick-release method to bring the pressure in the pot back to normal.

For medium-boiled eggs—Lock the lid onto the pot.

High pressure for 3 minutes. Close the lid and the pressure valve and then cook for 3 minutes. To get 3-minutes cook time, press STEAM button.

Pressure Release Use the quick-release method to bring the pot's pressure back to normal—but do not open the pot. Set the cooker aside, covered, for 1 minute. Use the quick-release method to bring the pot's pressure fully back to normal.

For hard-boiled eggs—Lock the lid onto the pot.

High pressure for 3 minutes. Close the lid and the pressure valve and then cook for 3 minutes. To get 3-minutes cook time, press STEAM button and use the TIME ADJUSTMENT button to adjust the cook time to 3 minutes.

Pressure Release. Turn off the machine or unplug it; set aside for 8 minutes. Use the quick-release method to bring the pot fully back to normal pressure.

For all eggs—Unlock and remove the lid. Transfer the eggs to a large bowl. Cut the top off a soft-boiled egg and serve it in an egg cup; peel the other kinds of eggs while still warm.

"Softboiled" Eggs

PREP: 5 MINUTES • PRESSURE: 3 MINUTES • TOTAL: 8 MINUTES • PRESSURE LEVEL: HIGH • RELEASE: QUICK
SERVES: 2

Ingredients

2 teaspoons unsalted butter, at room temperature, divided
2 large eggs
¼ teaspoon kosher salt, divided
Freshly ground black pepper
1-cup water, for steaming
2 slices of toast (optional)

Directions

1. **Preparing the Ingredients** Using ½ teaspoon of butter each, coat the insides of 2 heatproof custard cups or small ramekins. Crack 1 egg into each cup, and sprinkle each with ⅛ teaspoon of kosher salt and some pepper. Divide the remaining 1teaspoon of butter in half, and top each egg with one piece. (You can omit the butter on top of the egg, but it is delicious. Don't skip buttering the dish, though, or the egg won't come out.) Cover the cups with aluminum foil, crimping it down around the sides.
Add the water and insert the steamer basket or trivet. Carefully transfer the cups to the steamer insert.
2. **High pressure for 3 minutes** Close the lid and the pressure valve and then cook for 3 minutes. To get 3-minutes cook time, press STEAM button and use the TIME ADJUSTMENT button to adjust the cook time to 3 minutes.
3. **Pressure Release** Use the quick-release method.
4. **Finish the dish** Unlock but *don't remove* the lid for another 30 seconds; this will help ensure that the whites are fully cooked. Using tongs, remove the cups from the cooker and peel off the foil. Scoop each egg out onto a slice of toast (if desired).
5. Serve and Enjoy!

PER SERVING: CALORIES: 105; FAT: 9G; SODIUM: 388MG; CARBOHYDRATES: 0G; FIBER: 0G; PROTEIN: 3G

Grits with Cranberries Breakfast

PREP: 5 MINUTES • PRESSURE: 10 MINUTES • TOTAL: 15 MINUTES • PRESSURE LEVEL: HIGH • RELEASE: QUICK
SERVES: 4

Ingredients

¾ cup grits or polenta (not quick cook or instant)
3 cups water
⅛ teaspoon kosher salt
½ cup dried cranberries
1 tablespoon unsalted butter
1 tablespoon heavy (whipping) cream
2 tablespoons honey
½ cup slivered almonds, toasted

Directions

1. **Preparing the Ingredients.** In the Crock-Pot Multi-Cooker® combine the grits, water, kosher salt, and dried cranberries.
2. **High pressure for 10 minutes.** Lock the lid in place, and bring the cooker to high pressure by pressing Steam button and cook for 10 minutes.
3. **Pressure Release.** Use the quick-release method.
4. **Finish the dish.** Unlock and remove the lid. Quickly add the butter, heavy cream, and honey, and stir vigorously with a wooden spoon or paddle until smooth and creamy. Spoon into bowls, top with the toasted almonds, and serve.

PER SERVING: CALORIES: 251; FAT: 11G; SODIUM: 62MG; CARBOHYDRATES: 35G; FIBER: 3G; PROTEIN: 5G

Bacon and Onions Quiche

PREP: 5 MINUTES • PRESSURE: 8 MINUTES • TOTAL: 13 MINUTES • PRESSURE LEVEL: HIGH • RELEASE: QUICK
SERVES: 2

Ingredients

Butter, at room temperature, for coating
2 bacon slices, diced
¼ cup thinly sliced onion
¼ teaspoon kosher salt, plus additional for seasoning
2 large eggs
2 tablespoons whole milk
2 tablespoons heavy (whipping) cream
Freshly ground black or white pepper
1-cup water, for steaming

Directions

1. **Preparing the Ingredients.** Using a small amount of butter, coat the insides of 2 heatproof custard cups or small ramekins.
   Set the Crock-Pot Multi-Cooker® to "brown," add the bacon.
   Cook for 2 to 3 minutes, stirring occasionally, until the bacon renders most of its fat and is mostly crisp. Add the onion, and sprinkle with a pinch or two of kosher salt. Cook for about 3 minutes, stirring, until the onions just begin to brown. Transfer the bacon and onions to paper towels to drain briefly. Wipe out the inside of the Crock-Pot Multi-Cooker®. If you prefer, sauté the bacon and onions in a small skillet, and you won't have to clean out the Crock-Pot Multi-Cooker®.
   Into a small bowl, crack the eggs. Add the milk, heavy cream, and ¼ teaspoon of kosher salt, and season with the pepper. Whisk until the mixture is homogeneous; no streaks of egg white should remain. Pour one-quarter of the egg mixture into each cup or ramekin. Sprinkle half of the bacon and onions over each, and evenly divide the remaining egg over the bacon and onions.
   Add the water and insert the steamer basket or trivet. Carefully transfer the custard cups to the steamer insert. Place a sheet of aluminum foil over the cups. You don't have to crimp it down; it's just to keep steam from condensing on top of the custard.
2. **High pressure for 7 minutes.** Lock the lid in place, and bring the pot to high pressure, cook at high pressure for 7 minutes. To get 7-minutes cook time, press the Steam button and use the TIME ADJUSTMENT button to adjust the cook time to 7minutes.
3. **Pressure Release** Use the quick-release method.
4. **Finish the dish.** Unlock and remove the lid. Using tongs, carefully remove the custard cups from the Crock-Pot Multi-Cooker®. Cool for 1 to 2 minutes before serving. If you want to unmold the quiches, run the tip of a thin knife around the inside edge of the cups. One at a time, place a small plate over the top of the cups, and invert the quiches onto the plate.
5. Enjoy!

PER SERVING: CALORIES: 144; FAT: 12G; SODIUM: 392MG; CARBOHYDRATES: 3G; FIBER: 0G; PROTEIN

Chapter Seven

Main Dishes – Beef and Lamb

Classic Pot Roast

PREP: 5 MINUTES • PRESSURE: 90 MINUTES • TOTAL: 95 MINUTES • PRESSURE LEVEL: HIGH • RELEASE: QUICK AND NATURAL

SERVES: 6

Ingredients

1 tablespoon olive oil
One 3- to 3½-pound boneless beef chuck roast
1 teaspoon salt
½ teaspoon ground black pepper
1 large yellow onion, chopped
2 teaspoons minced garlic
Up to 1½ cups beef broth
3 tablespoons tomato paste
One 4-inch rosemary sprig
½ ounce dried mushrooms, preferably porcini
1½ pounds small white or yellow potatoes

Directions

1. **Preparing the Ingredients** Heat the oil in the Crock-Pot Multi-Cooker®. Turn on the pressure cooker to the Sauté setting then wait for it to boil.
   Season the roast with the salt and pepper; brown it on both sides, turning once, about 10 minutes. Transfer the meat to a large bowl.
   Add the onion; cook, stirring often, until translucent, about 4 minutes. Add the garlic; cook, stirring constantly, until aromatic, about 30 seconds. Pour 1¼ cup broth in the Crock-Pot Multi-Cooker®. Add the tomato paste and stir well until dissolved. Tuck the rosemary into the sauce and crumble in the mushrooms. Nestle the meat into the sauce, adding any juices in the bowl.
2. **High pressure for 60 minutes.** Close the lid and the pressure valve and then cook for 60 minutes. To get 60-minutes cook time, press Multigrain button and use the TIME ADJUSTMENT button to adjust the cook time to 60 minutes.
3. **Pressure Release** Use the quick-release method.
   Unlock and open the cooker; sprinkle the potatoes around the meat.
4. **High pressure for 30 minutes.** Close the lid and the pressure valve again and cook for 30 minutes. To get 30-minutes cook time, press Soup button.
5. **Pressure Release** Use the natural-release method -20 to 30 minutes.

6. **Finish the dish.** Transfer the roast to a cutting board; set aside for 5 minutes. Discard the rosemary sprig. Slice the meat into 2-inch irregular chunks and serve these in bowls with the vegetables, mushrooms, and broth.
7. Serve and Enjoy!

Mexican Beef

PREP: 10 MINUTES • PRESSURE: 35 MINUTES • TOTAL: 45 MINUTES • PRESSURE LEVEL: HIGH • RELEASE: NATURAL

SERVES 4-6

Ingredients

2½ pounds boneless beef short ribs, beef brisket, or beef chuck roast cut into 1½- to 2-inch cubes

1 tablespoon chili powder

1½ teaspoons kosher salt (Diamond Crystal brand)

1 tablespoon ghee or fat of choice

1 medium onion, thinly sliced

1 tablespoon tomato paste

6 garlic cloves, peeled and smashed

½ cup roasted tomato salsa

½ cup bone broth

½ teaspoon Red Boat Fish Sauce

Freshly ground black pepper

½ cup minced cilantro (optional)

2 radishes, thinly sliced (optional)

Directions

1. **Preparing the Ingredients:** Combine cubed beef, chili powder, and salt in a large bowl. Set the Crock Pot Express® to the "Brown/Sauté" Function, and add the ghee to the cooking insert. Once the fat's melted, add the onions and sauté until translucent. Stir in the tomato paste and garlic, and cook for 30 seconds or until fragrant. Toss in the seasoned beef and pour in the salsa, stock, and fish sauce.
2. **High pressure for 35 minutes**: Cover and lock the lid and Press the "Meat/Stew" button to switch to the pressure-cooking mode - 35 Minutes. When the stew is finished cooking, the CrockPot Express Multi-Cooker® will switch automatically to a "Keep Warm" mode.
3. **Pressure Release:** Use the Natural-release Method, -15 minutes.
4. **Finish the dish:** Unlock the lid and season to taste with salt and pepper.
5. Serve and Enjoy!

Per Serving Calories: 321.5; Saturated Fat: 2.8g; Fiber: 8.6g; Protein: 39.5g

Brisket With Veggies

PREP: 10 MINUTES • PRESSURE: 60 MINUTES • TOTAL: 70 MINUTES • PRESSURE LEVEL: HIGH • RELEASE: QUICK

SERVES 6

Ingredients

- 2 tbs. olive oil
- 5 or 6 red potatoes
- 2 lb. or larger regular brisket, rinsed and patted dry
- Fresh ground black pepper
- 3 tbs. heaping chopped garlic
- 1 lg. yellow onion
- 2 c. large chunks carrots
- 2-½ c. home made beef broth, or make from Knorr Beef Base
- 3 tbs. Worcestershire Sauce
- 4 bay leaves
- 5 or 6 red potatoes
- Granulated garlic
- Knorr Demi-Glace sauce
- ½ c. dehydrated onion
- 2 stalks celery in 1" chunks

Directions

1. **Preparing the Ingredients** Put the Crock-Pot Multi-Cooker® on the sauté setting. Put in 1 tbs. (more if needed) of the oil and caramelize the onions. Once golden, remove from pot, put in a bowl, and set aside. But keep the Crock-Pot Multi-Cooker® on the "Sauté" setting.

   Rub the freshly ground pepper on both sides of the brisket. Do the same with the granulated garlic. Add 1tbs. olive oil (or more) and only lightly sear the brisket on all sides.

   Add back the onions, garlic, Worcestershire sauce, bay leaves, dehydrated onion and beef broth.

2. **High pressure for 50 minutes**. Close the lid and the pressure valve and then cook for 50 minutes. To get 50-minutes cook time, press Meat/Stew button and use the TIME ADJUSTMENT button to adjust the cook time to 50 minutes.

   While the meat is cooking, peal and cut up all the veggies. When the meat is done, use the quick pressure release feature, and then remove the lid. Add all of the veggies, replace the lid and cook at high pressure for to 10 minutes. To get 10-minutes cook time, press Steam button

3. **Pressure Release** When the time is up, turn the pot off, use the quick release again, and remove the lid.

4. **Finish the dish.** Use a platter to remove the veggies and meat. Use the "Sauté" setting and bring the broth to a boil, then add the Knorr Demi-Glace mixing with a

Wisk. Adjust seasonings as needed. Serve with Cole Slaw or other salad, home made rolls or Italian garlic bread. Be sure to remove the bay leaves before serving.

5. Serve and Enjoy

Per Serving Calories: 425; Total Carbohydrates: 50g; Saturated Fat: 3.6g; Trans Fat: 0g; Fiber: 10.6g; Protein: 30.5g; Sodium: 490mg

Goulash

PREP: 5 MINUTES • PRESSURE: 25 MINUTES • PRESSURE LEVEL: HIGH • RELEASE: NATURAL
SERVES 6

Ingredients

    2 tablespoons olive oil, divided

    1½ pounds beef chuck roast, trimmed of excess fat and cut into 1½-inch cubes

    2 cups sliced onions

    2 garlic cloves, minced (about 2 teaspoons)

    Kosher salt, for seasoning

    ¼ cup sweet paprika

    2 teaspoons caraway seeds

    2 teaspoons dried marjoram or oregano

    3 cups Beef Stock or low-sodium broth

    1 (14-ounce) can diced tomatoes, drained

    2 large carrots, peeled and cut into 1-inch rounds (about 1½ cups)

    2 medium red bell peppers, cut into 1-inch pieces (about 1½ cups)

    ½ pound small red potatoes, left whole if less than 1½ inches in diameter, halved if larger

    Sour cream, for garnish (optional)

Directions

1. **Preparing the Ingredients:** Set the Crock-Pot Express® to "brown," heat 1 tablespoon of olive oil until it shimmers and flows like water. Add the beef, and sear on two sides, working in batches if necessary so as not to crowd the pan. Remove the beef from the cooker, and set aside. Add the remaining 1 tablespoon of olive oil to the pan; then add the onions and garlic, and sprinkle with a pinch or two of kosher salt. Cook, stirring, for about 3 minutes, or until the onions and garlic soften. Add the paprika, caraway seeds, and marjoram. Stir to coat the onions. Cook for about 1 minute, or until fragrant.
2. Pour the Beef Stock into the Crock-Pot Express®, and stir to dissolve the spices. Return the beef to the pot. Add the tomatoes, carrots, red bell peppers, and red potatoes.
3. Lock the lid in place, and bring the pot to high pressure.
1. **High pressure for 25 minutes**: Cook at high pressure for 25 minutes. To get 25-minutes cook time, press MEAT/STEW button and use the TIME ADJUSTMENT button to adjust the cook time to 25 minutes. When the timer goes off, turn the cooker off. ("warm" setting, turn off).
4. **Pressure Release.** After cooking, use the **natural method** to release pressure.
5. **Finish the dish.** Unlock and remove the lid. Let the goulash sit for 1 minute to allow any fat to rise to the surface. Spoon or blot off as much as possible.

Ground Beef Stew

PREP: 5 MINUTES • PRESSURE: 5 MINUTES • PRESSURE LEVEL: HIGH • RELEASE: QUICK

SERVES 4

Ingredients

- 1 tablespoon olive oil
- 1½ pounds lean ground beef (about 93% lean)
- 1 large yellow onion, chopped
- 1 large sweet potato (about 1 pound), peeled and shredded through the large holes of a box grater
- 1 teaspoon ground cinnamon
- 1 teaspoon ground cumin
- ½ teaspoon dried sage
- ½ teaspoon dried oregano
- ½ teaspoon salt
- ½ teaspoon ground black pepper
- 2 tablespoons yellow cornmeal
- 2 tablespoons honey
- 2½ cups beef broth

Directions

1. **Preparing the Ingredients**. Heat the oil in the Crock-Pot Express® turned to the "Brown/Sauté" function. Crumble in the ground beef; cook, stirring occasionally, until it loses its raw color and browns a bit, about 5 minutes. Add the onion; cook, stirring often, until softened, about 3 minutes. Stir in the sweet potato, cinnamon, cumin, sage, oregano, salt, and pepper. Cook for 1 minute, stirring constantly. Stir in the cornmeal and honey; cook for 1 minute, stirring often, to dissolve the cornmeal. Stir in the broth.
2. **High pressure for 5 minutes**. Lock the lid onto the pot. Switch the Crock-Pot Express® to cook at high pressure for 5 minutes. To get 5-minutes cook time, press STEAM button and use the TIME ADJUSTMENT button to adjust the cook time to 5 minutes.
3. **Pressure Release.** Use the quick-release method to drop the pot's pressure to normal.
4. **Finish the dish.** Unlock and open the lid. Stir well and set aside, loosely covered, for 5 minutes before serving.

Beef Bourguignon

PREP: 5 MINUTES • PRESSURE: 40 MINUTES • PRESSURE LEVEL: HIGH • RELEASE: NATURAL

SERVES 2

Ingredients

¾ pound beef chuck roast, trimmed of excess fat and cut into 2-inch chunks

½ teaspoon kosher salt

1 teaspoon unsalted butter

2 bacon slices, sliced crosswise into ½-inch pieces

2 teaspoons tomato paste

1 cup dry red wine (preferably Pinot Noir), divided

½ cup low-sodium beef broth

1 very small onion, cut into eighths

1 medium carrot, peeled and cut into ¼-inch slices (about ½ cup)

1 garlic clove, smashed

1 bay leaf

1 fresh thyme sprig

½ cup frozen pearl onions, thawed

½ cup "Sautéed" Mushrooms

1 tablespoon fresh minced parsley

Freshly ground black pepper

Directions

1. **Preparing the Ingredients**: Season the beef with the kosher salt. Set the Crock-Pot Express® to "brown," add the butter and bacon. Cook, stirring, for about 4 minutes, or until the bacon renders most of it's fat and is crisp. Remove the bacon, and set aside. Blot the beef chunks dry, and add them to the Crock-Pot Express®. Brown on all sides, about 10 minutes total, working in batches if necessary so as not to crowd the pan. Remove the beef from the pot, and set aside.

2. Add the tomato paste to the pot, and cook, stirring, for about 1 minute, or until the paste has darkened slightly. Add ½ cup of red wine, and cook, stirring, to release the browned bits from the bottom of the pan. Add the remaining ½ cup of red wine and the beef broth, onion, carrot, garlic, bay leaf, and thyme. Return the beef to the pot, and stir to combine.

3. **High pressure for 40 minutes**. Lock the lid in place, and bring the pot to high pressure. Cook at high pressure for 40 minutes. To get 40-minutes cook time, press MULTIGRAIN button.

4. **Pressure Release**. After cooking, use the natural method to release pressure.

5. **Finish the dish.** Unlock and remove the lid. Using tongs, remove the beef chunks to a bowl while you finish the sauce. Pour the sauce and vegetables through a strainer or colander into a fat separator. Discard the vegetables, bay leaf, and thyme sprig. When the fat has risen to the top of the separator, pour the defatted sauce back into

the cooker, and add the pearl onions. Turned to "brown," simmer the sauce for about 3 minutes, or until slightly thickened. Stir in the reserved bacon. Add the "Sautéed" Mushrooms and parsley, and season with pepper.

PER SERVING: CALORIES: 806; FAT: 50G; SODIUM: 869MG; CARBOHYDRATES: 16G;FIBER:3G;PROTEIN:48G

Meatballs With Artichokes

PREP: 10 MINUTES • PRESSURE: 8 MINUTES • PRESSURE LEVEL: HIGH • RELEASE: QUICK
SERVES 4

Ingredients

1½ pounds lean ground beef (preferably 93% lean)

½ cup dried orzo

1 medium shallot, peeled and shredded through the large holes of a box grater

1 tablespoon minced fresh dill fronds

2 teaspoons finely grated lemon zest

1 teaspoon minced garlic

1 large egg, at room temperature

2 tablespoons olive oil

One 28-ounce can diced tomatoes (about 3½ cups)

One 9-ounce box frozen artichoke heart quarters, thawed (about 2 cups)

½ cup rosé wine, such as Bandol

¼ cup loosely packed fresh basil leaves, minced

2 tablespoons loosely packed fresh oregano leaves, minced

½ teaspoon salt

½ teaspoon ground black pepper

Directions

1. **Preparing the Ingredients**. Mix the ground beef, orzo, shallot, dill, lemon zest, garlic, and egg in a large bowl until uniform. Form into twelve 2-inch balls. Heat the oil in the Crock-Pot Express® set to the "Brown/Sauté" function. Add the meatballs, just as many as will fit without crowding. Brown on all sides, turning occasionally, about 8 minutes. Transfer to a bowl and repeat with the rest of the meatballs.

2. Add the tomatoes, artichokes, wine, basil, oregano, salt, and pepper to the cooker; stir well to get any browned bits off the bottom of the pot. Return the meatballs and their juices to the sauce.

3. **High pressure for 8 minutes.**. Lock the lid onto the pot. Switch the Crock-Pot Express® to cook at high pressure for 8 minutes. To get 8-minutes cook time, press STEAM button and use the TIME ADJUSTMENT button to adjust the cook time to 8 minutes.

4. **Pressure Release**. Use the quick-release method to drop the pot's pressure back to normal.

5. **Finish the dish.** Unlock and open the pot. Stir gently before scooping the meatballs into serving bowls; ladle the sauce over them.

Shredded Barbecue Skirt Steak

PREP: 5 MINUTES • PRESSURE: 42 MINUTES • PRESSURE LEVEL: HIGH • RELEASE: NATURAL

SERVES 6

Ingredients

- ¼ cup unsweetened apple juice
- ¼ cup fresh lime juice
- Up to 3 canned chipotles in adobo sauce, stemmed, seeded, and chopped
- 1 medium shallot, chopped
- 1 tablespoon minced garlic
- 1 tablespoon packed fresh oregano leaves, finely chopped
- 1 tablespoon ground cumin
- ½ teaspoon salt
- ¼ teaspoon ground cloves
- 4 juniper berries (optional)
- 2 tablespoons rendered bacon fat
- 3 pounds beef skirt steak, cut into 6-inch-long pieces

Directions

1. **Preparing the Ingredients**. Place the apple juice, lime juice, chipotles, shallot, garlic, oregano, cumin, salt, cloves, and juniper berries, if using, in a large blender or food processor; cover and blend or process until smooth, stopping the machine a couple of times to scrape down the inside of the canister.
2. Melt the bacon fat in the Crock-Pot Express® turned to the "browning" function. Add one or two of the steaks and brown on both sides, about 4 minutes, turning once. Transfer to a plate and continue browning until you've worked your way through all the steak pieces.
3. Return the meat and any juices on its plate to the cooker. Pour the pureed sauce over the beef; stir well.
2. **High pressure for 42 minutes**. Lock the lid onto the pot. Set the pot to cook at high pressure for 42 minutes. To get 42-minutes cook time, press MULTIGRAIN button and use the TIME ADJUSTMENT button to adjust the cook time to 42 minutes.
4. **Pressure Release**. Turn off the Crock-Pot Express® or unplug it. Allow its pressure to fall to normal naturally, 15 to 20 minutes.
5. **Finish the dish.** Unlock and open the lid. Transfer the meat to a large cutting board; shred with two forks. Return the meat to the sauce in the cooker; stir well before serving.
6. Serve and Enjoy!

Easy Osso Bucco

PREP: 10 MINUTES • PRESSURE: 90 MINUTES • TOTAL: 100 MINUTES • PRESSURE LEVEL: HIGH • RELEASE: NORMAL

SERVES 4

Ingredients

4 veal or lamb shanks cut to size for the Crock-Pot Express®
¼ cup flour
½ tsp black pepper
½ tsp salt
½ tsp garlic powder
½ tsp onion powder
1 tsp thyme
1 tsp rosemary
¼ cup olive oil
1 Tbsp. butter
2 medium carrots chopped in large chunks
2 stalks celery cut into large chunks
1 medium to large onion chopped
2 cloves crushed garlic
1 to 2 cups chicken broth (keep in mind of the size of the Crock-Pot Express®)
2 lbs. red potatoes (washed)
2 Tbsp. butter

Directions

1. **Preparing the Ingredients**. Add the flour and the seasonings to a large bowl. Use a wire whisk to blend everything together.
2. Rinse the shanks and dry with a paper towel. Roll each shank in the flour mix and set aside on a plate Preheat a large skillet. Add the oil and bring to almost smoking. Place the shanks in the skillet and brown turning each shank to brown all sides of the shank. Once they are browned, set aside. Add the flour to the remaining oil and make a rue. Once the rue is made add the broth to loosen the rue into a sauce.
3. Pour ½ of the sauce on the Crock-Pot Express® and place each shank into the sauce standing upright. Fill in the gaps with the vegetables. Pour the remaining sauce over the shanks and vegetables.
3. **High pressure for 90 minutes.** Seal the Crock-Pot Express® and cook for approximately 90 minutes. To get 90-minutes cook time, press MEAT/STEW button and use the TIME ADJUSTMENT button to adjust the cook time to 90 minutes.
4. Towards the end of the cooking cycle, boil the red potatoes (skin on) until tender. Mash the potatoes adding 2 Tbsp. of butter. Salt and pepper to taste.
8. **Pressure Release**. . Turn off the Crock-Pot Express® or unplug it. Allow its pressure to fall to normal naturally, 15 to 20 minutes.

9. **Finish the dish.** Serve a lamb shank on a bed of potatoes. Add a large spoon of the vegetables. Ladle on some of the sauce from the cooker over the shank, vegetables and potatoes.

Per Serving Calories: 307.4; Carbohydrates: 5.6g; Fat: 8.7g; Fiber: 9.6g; Protein: 40.3g; Sodium: 840.6mg

One Pot Chinese Beef Stew

PREP: 10 MINUTES • PRESSURE: 30 MINUTES • TOTAL: 42 MINUTES • PRESSURE LEVEL: HIGH • RELEASE: NORMAL

SERVES 4-6

Ingredients

    1-2 Tsps. oil
    2 medium onions sliced
    ½ tsp sugar
    2 tsps. rice wine or sherry
    1 Tb soy sauce
    1 kg beef round, cubed into one inch pieces
    2 tsps. cornstarch
    Pinch of smoked Paprika
    1-2 tsp garlic powder
    Salt and pepper
    ½ cup broth, preferably beef
    1 Tb Worcestershire sauce
    1 can of mushrooms
    1-2 tsps. fresh ginger chopped finely
    1-2 tsps. cornstarch slurry if needed

Directions

1. **Preparing the Ingredients**. Place sugar, rice wine and soy sauce into Crock-Pot Express® using the "Brown/Sauté" mode fry for 30 seconds. Add beef broth and Worcestershire sauce, stir and close the lid.
4. **High pressure for 30 minutes**. Pressure cook on "Soup" for 30 minutes. To get 30-minutes cook time, press MEAT/STEW button and use the TIME ADJUSTMENT button to adjust the cook time to 30 minutes. Leave on keep warm for 3 minutes.
2. **Pressure Release**. Release pressure using the Natural Release Method.
3. **Finish the dish.** When meat is done, add chopped ginger, mushrooms (optional) and more salt and pepper (if needed). Sauté for another minute. Add cornstarch slurry to thicken to desired taste (if needed).
4. Serve with rice and stir fried greens or fresh cut veggies. Enjoy!

Corned Beef & Cabbage

PREP: 5 MINUTES • PRESSURE: 100 MINUTES • TOTAL: 125 MINUTES • PRESSURE LEVEL: HIGH • RELEASE: QUICK
SERVES 6

Ingredients

1 corned beef brisket (3-4 pounds)
4 cups water
1 small onion, peeled and quartered
3 garlic cloves, peeled and smashed
2 bay leaves
3 whole black peppercorns
½ teaspoon whole allspice berries
1 teaspoon dried thyme
1½ pounds small or medium red potatoes
5 medium carrots, peeled and cut into chunks
1 head cabbage, cut into wedges

Directions

1. **Preparing the Ingredients**. Place corned beef, water, onion quarters, garlic cloves, peppercorns, allspice, and thyme into the Crock-Pot Express®.
2. **High pressure for 90 minutes**. Lock lid in place and press MEAT/STEW button and use the TIME ADJUSTMENT button to adjust the cook time to 90 minutes.
3. **Pressure Release**. When cooking is complete, switch Crock-Pot Express® off and allow pressure to release naturally for 10 minutes, then quick release any remaining pressure.
4. **Finish the dish.** Remove the meat from the liquid and transfer to a plate. Cover with tin foil and allow to rest for 15 minutes while you prepare the vegetables. Add potatoes, carrots, and cabbage to liquid in Crock-Pot Express® and lock lid in place. Press "STEAM" button and set time for 10 minutes.
5. When cooking is complete, use the quick release pressure method. Use a slotted spoon to remove vegetables and serve with slices of corned beef, using some of the cooking liquid to moisten the meat and vegetables if necessary.

Lamb Curry

PREP: 10 MINUTES • PRESSURE: 40 MINUTES • TOTAL: 50 MINUTES • PRESSURE LEVEL: HIGH • RELEASE: NATURAL

SERVES 4

Ingredients

2 small onions
2 garlic cloves, peeled and smashed
1¼ cups plain yogurt
2 teaspoons kosher salt
1 tablespoon freshly squeezed lemon juice
1 tablespoon ground coriander
2 teaspoons ground cumin
1 teaspoon ground allspice
1½ teaspoon freshly ground black pepper
½ teaspoon ground ginger
2 tablespoons cornstarch
½ teaspoon red pepper flakes (optional)
1½ pounds boneless lamb shoulder, cut into 1½-inch cubes
Cooked rice or couscous, for serving
¼ cup chopped fresh mint

Directions

1. **Preparing the Ingredients**. Cut one of the onions into chunks. Place it, along all ingredients except for the lamb, rice, and mint, into a blender jar or food processor. Blend until mostly smooth. In a large bowl, pour the yogurt mixture over the lamb cubes. Stir to coat the meat evenly; then cover with plastic wrap or aluminum foil and marinate for 2 hours at room temperature, or in the refrigerator overnight.
2. Into the Crock-Pot Express®, pour the meat and marinade. Slice the remaining onion, and add it to the pot, stirring to combine.
3. **High pressure for 40 minutes**. Lock the lid in place, and bring the pot to high pressure for 40 minutes. To get 40-minutes cook time, press MEAT/STEW button and use the TIME ADJUSTMENT button to adjust the cook time to 40 minutes. When the timer goes off, turn the cooker off. ("Keep warm" setting, turn off).
4. **Pressure Release.** After cooking, use the natural method to release pressure.
5. **Finish the dish.** Unlock and remove the lid. Let the lamb sit for a few minutes to allow the fat to rise, and spoon off and discard the fat. Serve over rice or couscous, and garnish with the mint.

Braised Lamb Souvlaki

PREP: 5 MINUTES • PRESSURE: 40 MINUTES • TOTAL: 50 MINUTES • PRESSURE LEVEL: HIGH • RELEASE: QUICK
SERVES 4

Ingredients
1 lemon
1 tablespoon (2 g) finely chopped fresh rosemary
2 teaspoons paprika
2 teaspoons ground cumin
1/2 teaspoon ground coriander
3/4 teaspoon kosher salt, divided
3/4 teaspoon freshly ground black pepper, divided
1 1/2 pounds (680 g) boneless leg of lamb, trimmed and cut into 3-inch (7.5 cm) pieces
2 tablespoons (30 ml) extra-virgin olive oil, divided
2 cloves garlic, finely chopped
1 cup (235 ml) less-sodium chicken broth
2 tablespoons (32 g) tomato paste
1/4 small red onion, thinly sliced
2 plum tomatoes, cut into wedges
1/4 cup (25 g) pitted Kalamata olives, halved
1/4 cup (37 g) crumbled feta cheese
4 flatbreads, warmed
Plain yogurt, for serving

Directions

1. **Preparing the Ingredients**. Using a vegetable peeler, remove 3 strips of zest from the lemon and thinly slicecrosswise. Set aside. In a small bowl, combine the rosemary, paprika, cumin, coriander, 1/2 teaspoon of the salt, and 1/2 teaspoon of the pepper. Rub the lamb to coat with the mixture.
2. Turn the Crock-Pot Express® on to "Brown/Sauté". Heat 1 tablespoon (15 ml) of the olive oil. Add the lamb and cook until browned on both sides, about 5 minutes total. Add the garlic and cook, stirring, for 1 minute. Add the chicken broth, tomato paste, and lemon zest, and cook, stirring, for 1 minute.
3. **High pressure for 40 minutes**. Press [Cancel]. Lock the lid. Press MEAT/STEW button and use the TIME ADJUSTMENT button to adjust the cook time to 40 minutes.
4. **Pressure Release**. Use the "Quick Release" method to vent the steam, then open the lid. Using 2 forks, shred the lamb and toss it in the cooking liquid. Twenty minutes before the lamb is finished, place the onion in a medium bowl.
5. **Finish the dish.** Squeeze the juice of half the lemon on top and toss to combine. Let sit for 5 minutes. Add the tomatoes, olives, and the remaining 1 tablespoon (15 ml) olive oil, ¼ teaspoon salt, and 1/4 teaspoon pepper, and toss to combine. Fold in the feta.

6. Spread each flatbread with some yogurt, then top with the lamb and the tomato salad. Enjoy!

Lamb, Rice, And Chickpea Casserole

PREP: 10 MINUTES • PRESSURE: 30 MINUTES • TOTAL: 40 MINUTES • PRESSURE LEVEL: HIGH • RELEASE: QUICK

SERVES 6

Ingredients

2 pounds boneless leg of lamb, well trimmed and cut into 1½-inch pieces

1 medium yellow onion, halved

1 tablespoon salt

2 teaspoons whole allspice berries

2 teaspoons whole cloves

1 teaspoon black peppercorns

8 green cardamom pods

2 bay leaves

2 tablespoons olive oil

1 large yellow onion, halved and sliced into thin half-moons

1 tablespoon minced garlic

One 15-ounce can chickpeas, drained and rinsed (about 1¾ cups)

1 cup long-grain white rice, such as white basmati rice

½ teaspoon ground allspice

½ teaspoon ground ginger

Up to ½ teaspoon saffron threads

Directions

1. **Preparing the Ingredients**. Combine the lamb, onion halves, salt, allspice berries, cloves, peppercorns, cardamom pods, and bay leaves in the Crock-Pot Express®. Add enough tap water to cover all the ingredients.
2. **High pressure for 15 minutes**. Lock the lid onto the pot. Set the Crock-Pot Express® to cook at high pressure for 15 minutes. Press POULTRY button and use the TIME ADJUSTMENT button to adjust the cook time to 15 minutes.
3. **Pressure Release**: Use the quick-release method. Unlock and open the cooker. Cool for 5 minutes. Transfer the meat from the pot to a large bowl. Set a large bowl underneath a colander and drain the contents of the pot through the colander, catching the broth below. Discard the solids. Rinse out the cooker.
4. Turn the Crock-Pot Express® to its "Browning" mode. Add the oil, then the onion. Cook, stirring often, until softened, about 4 minutes. Stir the garlic; cook for just 30 seconds or so. Add the chickpeas, rice, allspice, ginger, and saffron; stir over the heat for 1 minute. Return the meat and any juices to the pot; pour in 2¼ cups of the reserved cooking liquid and stir well.
5. **High pressure for 15 minutes**. Lock the lid onto the pot again. Set the Crock-Pot Express® using the POULTRY button again to cook at high pressure for 15 minutes.
6. **Pressure Release**: Use the quick-release method to return the pot's pressure to normal, but do not remove the lid. Set the cooker aside for 5 minutes.
7. **Finish the dish.** Unlock and open the cooker. Stir well before serving. Enjoy!

Lamb Shanks Provençal

PREP: 10 MINUTES • PRESSURE: 40 MINUTES • TOTAL: 50 MINUTES • PRESSURE LEVEL: HIGH • RELEASE: NATURAL

SERVES 6

Ingredients

2 large (12-ounce) lamb shanks
1 teaspoon kosher salt, plus additional for seasoning
Freshly ground black pepper
1 tablespoon olive oil
1 cup sliced onion
2 garlic cloves, finely minced
2 medium plum tomatoes, coarsely chopped, or ½ cup diced canned tomatoes, drained
½ cup dry white wine or dry white vermouth
1 cup Chicken Stock or low-sodium broth
1 bay leaf
1 lemon, sliced very thin
⅓ cup pitted Kalamata olives
2 tablespoons coarsely chopped fresh parsley

Directions

Preparing the Ingredients.

1. Sprinkle the lamb shanks with 1 teaspoon of kosher salt and several grinds of pepper. The longer ahead of the cooking time you can do this, the better. Cover and let sit for 20 minutes to 2 hours at room temperature or refrigerate for up to 24 hours.
2. Turn the Crock-Pot Express® to "Brown/Sauté" heat the olive oil until it shimmers and flows like water. Add the lamb shanks, and brown on all sides, about 6 minutes total. Remove them to a plate. Add the onion and garlic, and sprinkle with a pinch or two of kosher salt. Cook, stirring, for about 3 minutes, or until the onions just begin to brown. Add the tomatoes, and cook until most of their liquid evaporates. Add the white wine, and stir, scraping up the browned bits from the bottom of the cooker.
3. Cook for 2 to 3 minutes, or until the wine reduces by about half; then add the Chicken Stock and bay leaf. Return the lamb shanks to the cooker, and place the lemon slices over them.

High pressure for 40 minutes.

4. Lock the lid in place, and bring the pot to high pressure. Cook at high pressure for 40 minutes. To get 40-minutes cook time, press MEAT/STEW button and use the TIME ADJUSTMENT button to adjust the cook time to 40 minutes.

Pressure Release.

5. After cooking, use the natural method to release pressure.

Finish the dish.

6. Unlock and remove the lid. Transfer the lamb to a cutting board or plate, and tent it with aluminum foil. Strain the sauce into a fat separator, and let it rest until the fat rises to the surface.
7. If you don't have a fat separator, let the sauce sit for a few minutes, then spoon or blot off any excess fat from the top and discard. Pour the defatted sauce back into the cooker along with the strained vegetables. If you want a thicker sauce, simmer the liquid for about 5 minutes, or until it reaches the desired consistency.
8. Stir in the olives and parsley. Place the shanks in shallow bowls, pour the sauce and vegetables over the lamb, and serve.
9. Lamb shanks benefit from salting in advance, which makes them much more flavorful and helps them brown beautifully. If you have the time, salt them up to 24 hours in advance. Place them on a tray and refrigerate, covered loosely with foil.

Korean Braised Short Ribs

PREP: 10 MINUTES • PRESSURE: 45 MINUTES • TOTAL: 55 MINUTES • PRESSURE LEVEL: HIGH • RELEASE: NATURAL

SERVES 4-6

Ingredients

1 teaspoon vegetable oil
2 green onions cut into 1-inch lengths
3 cloves garlic, smashed
3 quarter-sized slices of ginger
4 pounds beef short ribs, about 3 inches thick, cut into 3 rib portions
1/2-cup water
1/2-cup soy sauce
1/4-cup rice wine (or dry sherry)
1/4-cup pear juice (or apple juice)
2 teaspoons sesame oil
Minced green onions
Gochujang sauce

Directions

1. **Preparing the Ingredients** Heat the vegetable oil in the Crock-Pot Multi-Cooker® using the "Sauté" function, until the oil is shimmering. Add the green onion, garlic, and ginger, and sauté for 1 minute, or until you can smell garlic. Add the short ribs, water, soy sauce, rice wine, pear juice and sesame oil. Stir until the ribs are completely coated.
2. **High pressure for 45 minutes**. Lock the lid on the Crock-Pot Multi-Cooker® and then cook for 45 minutes. To get 45-minutes cook time, press Meat/Stew button and use the TIME ADJUSTMENT button to adjust the cook time to 45 minutes.
3. **Pressure Release** Let the pressure to come down naturally for at least 15 minutes, then quick release any pressure left in the pot.
4. **Finish the dish** Remove the short ribs from the pot with a slotted spoon.
5. Serve the ribs with the degreased sauce.

Beef Ribs

PREP: 10 MINUTES • PRESSURE: 60 MINUTES • TOTAL: 70 MINUTES • PRESSURE LEVEL: HIGH • RELEASE: NORMAL

SERVES 4-6

Ingredients

- 1 tablespoon sesame oil
- 2 cloves garlic, peeled and smashed
- 1" knob fresh ginger, peeled and finely chopped
- 1 pinch red pepper flakes
- ¼ cup rice vinegar (or white balsamic vinegar)
- ⅓ cup raw sugar
- ⅔ cup soy sauce
- ⅔ cup salt-free (home made) beef stock
- 4 pounds (2k) beef ribs (about 8), ask butcher to saw or chop them in half
- 2 tablespoons cornstarch
- 1-2 tablespoons water

Directions

1. **Preparing the Ingredients** Turn on the Crock-Pot Multi-Cooker® to "Sauté" mode.
   Add sesame oil garlic, ginger and red pepper flakes and sauté for a minute.
   Then, de-glaze with vinegar, mix-in the sugar, soy sauce and beef stock - mix well.
   Add the ribs to the Crock-Pot Multi-Cooker® coating them with the mixture.
2. **High pressure for 60 minutes**. Close and lock the lid of the Crock-Pot Multi-Cooker®, cook at high pressure for 60 minutes. To get 60-minutes cook time, press Meat/Stew button and use the TIME ADJUSTMENT button to adjust the cook time to 60 minutes.
3. **Pressure Release** Use the Natural release method (20 minutes).
4. **Finish the dish** Remove the ribs, and place on a cookie sheet and slide under the broiler for about 5 minutes to brown. Make a slurry with the corn starch and water and then mix into the rib cooking liquid in the Crock-Pot Multi-Cooker®. "Sauté" the mixture until it reaches the desired consistency.
5. Serve and Enjoy!

Per Serving Calories: 307.3; Carbohydrates: 8.6g; Fat: 10.7g; Fiber: 10.6g; Protein: 32.3g; Sodium: 1654.6mg; Cholesterol: 89.2g

Pulled BBQ Beef Sandwiches

PREP: 10 MINUTES • PRESSURE: 35 MINUTES • TOTAL: 45 MINUTES • PRESSURE LEVEL: HIGH • RELEASE: NORMAL

SERVES 2-4

Ingredients

2 pounds – Beef of choice

2 cps – Water

4 cps – Finely shredded Cabbage (the secret ingredient and you'll never know it's in there.)

1/2 cup – Of your favorite BBQ Sauce

1 cup – Ketchup

1/3 cup – Worcestershire Sauce

1 tblsp – Horse Radish

1 tblsp – mustard

Directions

1. **Preparing the Ingredients.** Add and stir in ingredients to your Crock-Pot Multi-Cooker®.
2. **High pressure for 35 minutes.** Lock the lid on the Crock-Pot Multi-Cooker ® and then cook for 35 minutes. To get 35-minutes cook time, press Meat/Stew button.
3. **Pressure Release.** Use natural release method.
4. **Finish the dish** Set the beef aside. Set the Crock-Pot Multi-Cooker® to a "Sauté" mode, Sauté the sauce until it reaches the desired consistency.
5. Serve and Enjoy.

Main Dishes – Chicken And Turkey

Easy Chicken

PREP: 5 MINUTES • PRESSURE: 15 MINUTES • TOTAL: 20 MINUTES • PRESSURE LEVEL: HIGH • RELEASE: QUICK
SERVES 4

Ingredients

    1 lb. boneless skinless chicken breasts, frozen
    1/2 cup water
    1/2 cup flavorful liquid of your choice

Directions

1. **Preparing the Ingredients**. In a measuring cup or small bowl, mix together the water and flavorful liquid of your choice. Place the frozen chicken in the Crock-Pot Express® liner, and pour the liquid over the chicken.
2. **High pressure for 15 minutes**. Close the lid, press the 'Poultry' button. For standard chicken breasts (~4-6 oz. each), cook for 15 minutes; for extra-large chicken breasts (~1 lb. each), cook for 30 minutes.
3. **Pressure Release**. Use the quick release method.
4. **Finish the dish.** Transfer the chicken breasts to a plate and shred into bite-sized pieces with two forks. While you shred the chicken, you can optionally turn on Crock-Pot Express's 'Brown/Sauté' mode to reduce the sauce if it is too thin for your taste. Return the shredded chicken to the sauce and toss to coat.
5. Serve and Enjoy!

Chicken Leg Quarters With Rosemary

PREP: 5 MINUTES • PRESSURE: 18 MINUTES • TOTAL: 23 MINUTES • PRESSURE LEVEL: HIGH • RELEASE: NATURAL SERVES 4 TO 6

Ingredients

    2 tablespoons olive oil

    2 tablespoons loosely packed fresh rosemary leaves, minced

    1 tablespoon mild paprika

    1 teaspoon salt

    ½ teaspoon ground black pepper

    4 chicken leg-and-thigh quarters (3–3½ pounds total weight), skin removed

    ¾ cup chicken broth

    8 medium garlic cloves

Directions

1. **Preparing the Ingredients**. Make a paste from the olive oil, rosemary, paprika, salt, and pepper by stirring it in a small bowl with a fork. Rub this paste into the quarters.
2. Pour the broth into the Crock-Pot Express®; set the quarters in the pot, overlapping only as necessary. Tuck the garlic cloves around the quarters.
3. **High pressure for 18 minutes**. Lock the lid onto the pot. Set the Crock-Pot Express® to cook at high pressure for 18 minutes. To get 18-minutes cook time, press Poultry button and use the TIME ADJUSTMENT button to adjust the cook time to 18 minutes.
4. **Pressure Release**. Use the natural-release method, 12 to 15 minutes.
5. **Finish the dish.** Unlock and open the pot. Transfer the chicken to serving plates; stir the sauce and spoon over the meat.

Curried Lemon Coconut Chicken

PREP: 5 MINUTES • PRESSURE: 35 MINUTES • TOTAL: 40 MINUTES • PRESSURE LEVEL: HIGH • RELEASE: QUICK

SERVES 6

Ingredients

1 can full fat coconut milk
¼ c. lemon juice
1 Tbs. curry powder
1 tsp. turmeric
½ tsp. salt
About 4 lbs. chicken - breasts, thighs, or a combo (whatever you have)
Optional: ½-1 tsp. lemon zest

Directions

1. **Preparing the Ingredients**. Mix the coconut milk, lemon juice and spices together in a bowl or glass measuring cup. Pour a little bit on the bottom of the Crock-Pot Express®. Add the chicken. Pour in the rest, including the coconut cream chunk if you've got one, on top of the chicken.
2. **High pressure for 25 minutes**. Lock in the lid and close the valve. Turn the Crock-Pot Express® to "Poultry" which should be 15 minutes at high pressure. If working with frozen chicken breasts, add 10 minutes to the cook time and you should be fine.
3. **Pressure Release**. Use the quick-release method. Test chicken for doneness by cutting open and observing the center.
4. **Finish the dish.** Use 2 forks to shred the chicken up in the pot. Add ½-1 tsp. lemon zest after cooking. Serve with steamed or roasted veggies or over rice.
5. Enjoy!

Per Serving Calories: 351; Total Carbohydrates: 35g; Saturated Fat: 3.8g; Fiber: 9.6g; Protein: 29.5g

Chicken with Soy Sauce

PREP: 10 MINUTES • PRESSURE: 30 MINUTES • TOTAL: 40 MINUTES • PRESSURE LEVEL: LOW • RELEASE: NATURAL

SERVES 8

Ingredients

  1 medium size chicken
  1 green onion minced
  1 small piece of ginger minced
  2 tablespoons sugar
  2 teaspoons salt
  2 teaspoons soy sauce
  1 tablespoon cooking wine or 2 tablespoons wine

Directions

1. **Preparing the Ingredients**: Mix the 2 teaspoon salt, 2 tablespoons sugar and seasoned the outside/inside chicken with them evenly. Add one teaspoon salt and cover the bottom of the Crock-Pot Express t®. Put seasoned chicken, soy sauce, wine into Crock-Pot Express®.
2. **High pressure for 30 minutes**. Press the "Poultry" button and wait till it is done. Turn the chicken over then press the "Poultry" button again.
3. **Pressure Release**. Use the quick release method.
4. **Finish the dish.** Then the delicious chicken, retaining flavor and nutrition, is ready to be served. Cut the chicken into pieces. Mix ginger and green onion with chicken oil to make a dipping sauce.
5. Serve and Enjoy!

Chicken Congee

PREP: 5 MINUTES • PRESSURE: 60 MINUTES • TOTAL: 65 MINUTES • PRESSURE LEVEL: HIGH • RELEASE: NATURAL & QUICK

SERVES 7 Cups

Ingredients

- 1 rice measuring cup (180 ml) Jasmine rice
- 7 cups water (using standard 250 ml cup)
- 5 – 6 chicken drumsticks
- 1 tablespoon ginger, sliced into strips
- Green onions for garnish
- Salt to taste

Directions

1. **Preparing the Ingredients**. Rinse rice in the pot under cold water by gently scrubbing the rice with your fingertips in a circling motion. Pour out the milky water, and continue to rinse until water is clear. Drain well. Add 7 cups of water (using standard 250 ml cup) and ginger into the pot.
2. **High pressure for 15 minutes**. Close lid, press Pultry button, and cook at high pressure for 15 minutes in the Crock-Pot Express®.
3. **Pressure Release.** Natural release for 5 minutes, then Quick Release. Be careful as you do quick release.
4. Open the lid, Do Not Stir. Add the chicken drumsticks into the Crock-Pot Express®, then close the lid again.
5. **High pressure for 20 minutes** Cook at high pressure for another 20 minutes, To get 20-minutes cook time, press Poultry button and use the TIME ADJUSTMENT button to adjust the cook time to 20 minutes.
6. **Pressure Release**. use natural release for 15 minutes, and quick release after.
7. **Finish the dish.** Heat up the pot (press "Brown/Sauté" button), season with salt and stir until desired consistency. Use tongs and fork to separate the chicken meat from the bones (they literally fall off the bone) and remove the chicken bones. Remove congee from heat and garnish with green onions.
8. Serve and Enjoy!.

Thai Lime Chicken

PREP: 10 MINUTES • PRESSURE: 10 MINUTES • TOTAL: 20 MINUTES • PRESSURE LEVEL: LOW • RELEASE: QUICK

SERVES 4-6

Ingredients

2 lbs. boneless skinless chicken thighs
1 c lime juice
1/2 c fish sauce
1/4 c olive oil
2 T coconut nectar
1 t grated fresh ginger
1 t chopped fresh mint
2 t chopped fresh cilantro

Directions

1. **Preparing the Ingredients**. Place the Chicken in bottom of the Crock-Pot Express®. Combine all remaining ingredients in a mason jar and shake well. Pour over chicken.
2. **High pressure for 10 minutes**. Select "poultry" Crock-Pot Express® setting and reduce time to 10 minutes.
3. **Pressure Release.** Use quick-release method.
4. Serve and Enjoy!

Chicken with Mushrooms

PREP: 10 MINUTES • PRESSURE: 20 MINUTES • TOTAL: 30 MINUTES • PRESSURE LEVEL: LOW • RELEASE: NATURAL

SERVES 4

Ingredients

- ½ cup flour (all-purpose)
- ½ tsp salt
- ½ tsp pepper
- 6 boneless skinless chicken (cut into bite-sized chunks)
- 2 tbsp. oil
- 1 onion minced
- 1 (10-ounce) can tomato sauce
- 1 tsp vinegar
- 1 (4-ounce can – I used fresh) sliced mushrooms
- 1 tbsp. sugar
- 1 tsp garlic – minced
- 1 tbsp. dried oregano
- 1 tsp dried basil
- 1 tsp chicken bouillon granules
- 1 cup Romano cheese

Directions

1. **Preparing the Ingredients**. Turn your Crock-Pot Express® onto the "Brown/Sauté" feature and place the chicken in oil until chicken starts to brown. Add onion and garlic and cook until they start to become translucent. Add remaining ingredients except Romano cheese. Stir to combine ingredients.
5. **High pressure for 10 minutes**. Close the Crock-Pot Express® lid. Select "poultry" Crock-Pot Express® setting and reduce time to 10 minutes. After cooking, use "keep warm" mode for 10 minutes.
6. **Pressure Release**. Use Quick release Method.
7. **Finish the dish.** Remove lid and add Romano cheese and stir. Add the butter-flour paste-mixture to thicken sauce.
2. Serve and Enjoy!

Per Serving Calories: 310; Total Carbohydrates: 40g; Saturated Fat: 3.8g; Fiber: 9.6g; Protein: 28.5g; Sodium: 494mg

Filipino Chicken Adobo

PREP: 10 MINUTES • PRESSURE: 15 MINUTES • TOTAL: 25 MINUTES • PRESSURE LEVEL: LOW • RELEASE: NATURAL

SERVES 8

Ingredients

- 4 -5 lbs. chicken thighs
- 1/2 cup white vinegar
- 1/2 cup soy sauce
- 4 cloves garlic, crushed
- 1 tsp. black peppercorns
- 3 bay leaves

Directions

1. **Preparing the Ingredients**. Set your Crock-Pot Express® to the Poultry setting and add the chicken and all ingredients. You do not have to Sauté anything.
2. **High pressure for 15 minutes** Close the lid, and cook for 15 minutes.
3. **Pressure Release**. Use Quick release Method.
4. **Finish the dish**. Serve with white Jasmine rice, Calrose white rice.

Per Serving Calories: 251; Total Carbohydrates: 35g; Saturated Fat: 2.8g; Fiber: 2.6g; Protein: 28.5g; Sodium: 394mg; Cholesterol: 52.5mg

Chicken Piccata

PREP: 15 MINUTES • PRESSURE: 30 MINUTES • TOTAL: 45 MINUTES • PRESSURE LEVEL: LOW • RELEASE: NORMAL

SERVES 4-6

Ingredients

1½ pounds boneless chicken breasts, trimmed (6 breasts)
½ cup all purpose flour plus 1 tablespoon, divided
2 teaspoon kosher salt, divided
2 tablespoons olive oil
3 cloves garlic, minced
¾ cup chicken broth
⅓ cup fresh lemon juice
2 tablespoons cooking sherry
1 teaspoon dried basil
1 teaspoon dried oregano
3 ounces capers, drained
¼ cup sour cream
Lemon slices for garnish

Directions

1. **Preparing the Ingredients**. Place ½ cup flour and 1 teaspoon salt in a large zip-top bag. Place the chicken breasts in the bag and shake to coat in flour. Shake off any excess flour and transfer to a plate. Heat your Crock-Pot Express® to Brown/Sauté for 30 minutes. When the pot comes to temperature coat the bottom of the pot with olive oil. Working 2 pieces at a time brown the chicken breasts on both sides. When browned, transfer to a clean plate.
2. When all the chicken is browned add the garlic to the pot and cook until softened (about 1 minute) stirring constantly. Add in the chicken broth, lemon juice, sherry, basil and oregano. Add the chicken back to the pot and nestle into the liquid. Top with the capers.
6. **High pressure for 10 minutes**. Place the lid on the pot and set to pressure cook on high for 10 minutes. To get 10-minutes cook time, press "Poultry button and use the TIME ADJUSTMENT button to adjust the cook time to 10 minutes. While the chicken is cooking mix together the sour cream and remaining 1 tablespoon of flour in a separate bowl. Set aside.
7. **Pressure Release**. When the 10 minutes is up release the pressure, using the Quick method, and remove the chicken from the pot placing on a serving platter.
3. **Finish the dish**. Whisk in the sour cream to the liquid and cook for 1 minute. The liquid will be very hot, you don't need to heat the pot.
4. Pour the sauce over the chicken and serve warm over pasta if desired.
5. Enjoy!

Fall-Off-The-Bone Pressure Cooker Chicken

PREP: 10 MINUTES • PRESSURE: 35 MINUTES • TOTAL: 45 MINUTES • PRESSURE LEVEL: HIGH • RELEASE: NORMAL

SERVES 10

Ingredients

    1 whole - 4lb. organic chicken
    1 Tbsp. Organic Virgin Coconut Oil
    1 tsp. paprika
    1½ cups Pacific Organic Bone Broth (Chicken)
    1 tsp. dried thyme
    ¼ tsp. freshly ground black pepper
    2 Tbsp. lemon juice
    ½ tsp. sea salt
    6 cloves garlic, peeled

Directions

1. **Preparing the Ingredients**. In a small bowl, combine paprika, thyme, salt, and pepper. Rub seasoning over outside of bird. Heat oil in the Crock-Pot Express® to shimmering. Add chicken, breast side down and cook 6-7 minutes.
2. Flip the chicken and add broth, lemon juice and garlic cloves.
3. **High pressure for 25 minutes**. Lock Crock-Pot Express® lid and set for 25 minutes on high. To get 25-minutes cook time, press "Poultry" button and use the TIME ADJUSTMENT button to adjust the cook time to 25 minutes.
4. **Pressure Release**. Let the Crock-Pot Express® release naturally.
5. **Finish the dish**. Remove from Crock-Pot Express® and let stand for 5 minutes before carving.
6. Serve.

Easy Turkey Drumsticks

PREP: 5 MINUTES • PRESSURE: 40 MINUTES • TOTAL: 45 MINUTES • PRESSURE LEVEL: HIGH • RELEASE: NATURAL

SERVES 6

Ingredients

6 turkey drumsticks
1 tablespoon Diamond Crystal kosher salt
2 packed teaspoons brown sugar
1 teaspoon fresh ground black pepper
1/2 teaspoon garlic powder
1/2 cup water
1/2 cup soy sauce

Directions

1. **Preparing the Ingredients**. Mix the salt, brown sugar, pepper, and garlic powder, breaking up any clumps of brown sugar. Sprinkle evenly over the turkey drumsticks. Pour the water and soy sauce into the Crock-Pot Express®, then add the drumsticks.
7. **High pressure for 25 minutes**. Lock the lid and cook at high pressure for 25 minutes. To get 25-minutes cook time, press "Poultry" button and use the TIME ADJUSTMENT button to adjust the cook time to 25 minutes.
2. **Pressure Release**. Use natural-release method for 15 minutes, and then use quick-release.
3. **Finish the dish.** Lift the drumsticks out of the pot with tongs - be careful, they are fall-apart tender. Let the fat float to the top, and pass the de-fatted liquid at the table as a sauce.
4. Serve and Enjoy!

Turkey Legs

PREP: 10 MINUTES • PRESSURE: 30 MINUTES • TOTAL: 40 MINUTES • PRESSURE LEVEL: HIGH • RELEASE: NATURAL

SERVES 2-4

Ingredients

- 2 turkey legs
- 1 tablespoon olive oil
- 1 small onion, sliced
- 3 cloves garlic, roughly minced
- 1 celery stalk, chopped
- 2 bay leaves
- A pinch of rosemary
- A pinch of thyme
- A dash of sherry wine
- 1 cup unsalted homemade chicken stock
- 1 tablespoon light soy sauce
- Kosher salt and ground black pepper to taste

Directions

1. **Preparing the Ingredients**. Season the turkey legs with generous amount of kosher salt and ground black pepper. Heat up your Crock-Pot Express®, press "Brown/Sauté" button. Add 1 tablespoon of olive oil into the pot. Ensure to coat the oil over the whole bottom of the pot. Add the seasoned turkey legs into the pot, then let it brown for roughly 2 - 3 minutes per side. Remove and set aside.
2. Add the sliced onion and stir. Add a pinch of kosher salt and ground black pepper to season if you like. Cook the onions for roughly one minute until soften. Add garlic, and then stir for 30 seconds until fragrance. Add in chopped celery and cook for roughly one minute.
3. Add in a dash of sherry wine, deglaze the bottom of the pot with a wooden spoon. Allow it to cook for a moment for the alcohol to evaporate. Add chicken stock and light soy sauce, then taste the seasoning. Add in more salt and pepper if desired.
8. **High pressure for 20 minutes**. Place the turkey legs into the Crock-Pot Express®, then close lid. Pressure cook at high pressure for 20 minutes. To get 20-minutes cook time, press "Poultry" button and use the TIME ADJUSTMENT button to adjust the cook time to 20 minutes.
4. **Pressure Release.** Use Natural-release method for 10 minutes and then quick-release.
5. Serve and Enjoy!

Turkey Breast

PREP: 20 MINUTES • PRESSURE: 30 MINUTES • TOTAL: 50 MINUTES • PRESSURE LEVEL: HIGH • RELEASE: NATURAL & QUICK

SERVES 4

Ingredients

6.5 lb. bone-in, skin-on turkey breast
Salt and pepper, to taste
1 (14 oz.) can turkey or chicken broth
1 large onion, quartered
1 stock celery, cut in large pieces
1 sprig thyme
3 tablespoons cornstarch
3 tablespoons cold water

Directions

1. **Preparing the Ingredients**. Season turkey breast liberally with salt and pepper. Put trivet in the bottom. Add chicken broth, onion, celery and thyme. Add the turkey to the cooking pot breast side up.
2. **High pressure for 30 minutes.** Lock lid in place, select High Pressure and 30 minutes cooking time. To get 20-minutes cook time, press "Poultry" button and use the TIME ADJUSTMENT button to adjust the cook time to 20 minutes.
3. **Pressure Release**. Use a natural pressure release for 10 minutes, then do a quick pressure release. Check if the turkey is done. If it isn't, lock the lid in place and cook it for a few more minutes.
4. **Finish the dish.** Carefully remove turkey and place on large plate. Cover with foil. Strain and skim the fat off the broth. Whisk together corn starch and cold water; add to broth in cooking pot. Select Sauté and stir until broth thickens. Add salt and pepper to taste. Slice the turkey and serve immediately.
5. Enjoy!

Cranberry Braised Turkey Wings

PREP: 10 MINUTES • PRESSURE: 25 MINUTES • TOTAL: 35 MINUTES • PRESSURE LEVEL: HIGH • RELEASE: NATURAL

SERVES 2-4

Ingredients

- 2 Tbsp. Butter
- 2 Tbsp. Oil
- 4 Turkey wings (2-3 lbs.)
- Salt and Pepper, to taste
- 1 cup Dry Cranberries or "Crasins" (soaked in boiling water for 5 minutes) or 1½ cup Fresh Cranberries or 1 cup of canned cranberries, rinsed
- 1 med Onion, roughly sliced
- 1 cup shelled Walnuts
- 1 cup Freshly Squeezed Orange juice (or prepared juice with no added sugar)
- 1 bunch Fresh Thyme

Directions

1. **Preparing the Ingredients** Set the Crock-Pot Express® to "Brown/Sauté", melt the butter and swirl the olive oil. Brown the turkey wings on both sides adding salt and pepper to taste. Make sure that the skin side is nicely colored. Remove the wings briefly from the Crock-Pot Express® and add the onion, then on top of that add the wings, cranberries, walnuts, a little bundle of Thyme. Pour the orange juice over the turkey.
2. **High pressure for 20 minutes**. Close and lock the Crock-Pot Express®. Cook for 20 minutes at high pressure. To get 20-minutes cook time, press "Poultry" button and use the TIME ADJUSTMENT button to adjust the cook time to 20 minutes.
3. **Pressure Release**. Use the Natural method.
4. **Finish the dish.** Remove the thyme bundle and carefully remove the wings to a serving dish. Slide the serving dish under the broiler for about 5 minutes or until the wings are sufficiently caramelized. In the meantime, reduce the cooking liquid to about half. Pour the reduced liquid, walnuts, onions and cranberries over the wings and serve.
5. Enjoy!

Per Serving Calories: 240.3; Fat: 4.8g; Saturated Fat: 1.9g; Carbohydrates: 18.8g; Sodium: 746.8mg; Fiber: 2.5g; Protien: 20.6g; Cholesterol: 32.9mg

Paleo Turkey and Gluten Free Gravy

PREP: 10 MINUTES • PRESSURE: 35 MINUTES • TOTAL: 75 MINUTES • PRESSURE LEVEL: HIGH • RELEASE: QUICK

SERVES 6

Ingredients

1 4-5 pound bone-in, skin-on turkey breast
Salt
Black pepper (omit for AIP)
2 tablespoons ghee or butter (use coconut oil for AIP)
1 medium onion, cut into medium dice
1 large carrot, cut into medium dice
1 celery rib, cut into medium dice
1 garlic clove, peeled and smashed
2 teaspoons dried sage
¼ cup dry white wine
1½ cups bone broth (preferably from chicken or turkey bones)
1 bay leaf
1 tablespoon tapioca starch (optional)

Directions

1. **Preparing the Ingredients**. Set the "Brown/Sauté" function. Pat turkey breast dry and generously season with salt and pepper. Melt cooking fat in the Crock-Pot Express®. Brown turkey breast, skin side down, about 5 minutes, and transfer to a plate, leaving fat in pot. Add onion, carrot, and celery to pot and cook until softened, about 5 minutes. Stir in garlic and sage and cook until fragrant, about 30 seconds.
2. Pour in wine and cook until slightly reduced, about 3 minutes. Stir in broth and bay leaf. Using wooden spoon, scrape up all browned bits stuck on bottom of pot. Place turkey skin side up in post with any accumulated juices.
6. **High pressure for 35 minutes**. Lock lid in place and set the Crock-Pot Express® for 35 minutes on high pressure. To get 35-minutes cook time, press "Poultry" button and use the TIME ADJUSTMENT button to adjust the cook time to 35 minutes.
3. **Pressure Release**. Use quick-release method and carefully remove lid.
4. **Finish the dish.** Transfer turkey breast to carving board or plate and tent loosely with foil, allowing it to rest while you prepare the gravy. Use an immersion blender or carefully transfer cooking liquid and vegetables to blender and puree until smooth. Return to medium high heat and cook until thickened and reduced to about 2 cups. Adjust seasoning to taste. Slice turkey breast and serve with hot gravy.
5. Enjoy!

Main Dishes – Pork

Pulled Pork

PREP: 5 MINUTES • PRESSURE: 80 MINUTES • TOTAL: 60 MINUTES • PRESSURE LEVEL: HIGH • RELEASE: NATURAL
SERVES 8

Ingredients
2 tablespoons smoked paprika
2 tablespoons packed dark brown sugar
1 tablespoon ground cumin
2 teaspoons ground black pepper
½ tablespoon dry mustard
1 teaspoon ground coriander
1 teaspoon dried thyme
1 teaspoon onion powder
1 teaspoon salt
½ teaspoon garlic powder
½ teaspoon ground cloves
½ teaspoon ground cinnamon
One 4- to 4½-pound bone-in skinless pork shoulder, preferably pork butt
Up to 1½ cups light-colored beer, preferably a pale ale or amber lager

Directions

1. **Preparing the Ingredients.** Mix the smoked paprika, brown sugar, cumin, pepper, mustard, coriander, thyme, onion powder, salt, garlic powder, cloves, and cinnamon in a small bowl. Massage the mixture all over the pork.
1. Set the pork Crock-Pot Express®. Pour 1 cup beer into the electric model without knocking the spices off the meat.
2. **High pressure for 50 minutes**. Lock the lid onto the pot. Cook at high pressure for 80 minutes. To get 80-minutes cook time, press "Meat/Stew" button and use the TIME ADJUSTMENT button to adjust the cook time to 80 minutes.
3. **Pressure Release** Let its pressure fall to normal naturally, 25 to 35 minutes.
4. **Finish the dish.** Transfer the meat to a large cutting board. Let stand for 5 minutes. Use a spoon to skim as much fat off the sauce in the pot as possible. Set the "brown/sauté "function. Bring the sauce to a simmer, stirring occasionally; continue boiling the sauce, stirring often, until reduced by half, 7 to 10 minutes. Use two forks to shred the meat off the bones; discard the bones and any attached cartilage. Pull any large chunks of meat apart with the forks and stir the meat back into the simmering sauce to reheat.
5. Serve and Enjoy!

Pork with Apple Juice

PREP: 10 MINUTES • PRESSURE: 20 MINUTES • TOTAL: 40 MINUTES • PRESSURE LEVEL: HIGH • RELEASE: NATURAL

SERVES 3

Ingredients

1 tablespoon vegetable oil
4 (6 ounce) pork tenderloins
1 (16 ounce) package sauerkraut, drained
1 cup water
6 fluid ounces apple juice
2 teaspoons fennel seed
12 red new potatoes, halved

Directions

1. **Preparing the Ingredients**. Heat oil in the Crock-Pot Express®, using Brown/Sauté button; brown pork tenderloins in the hot oil, about 5 minutes per side. Distribute sauerkraut around the pieces of pork and pour in water and apple juice; sprinkle with fennel seeds.
2. **High pressure for 15 minutes**. Cover the Crock-Pot Express® and cook on High Pressure for 15 minutes. To get 15-minutes cook time, press "Meat/Stew" button and use the TIME ADJUSTMENT button to adjust the cook time to 15 minutes.
3. **Pressure Release**. Use Natural release Method.
4. **High pressure for 5 minutes** Place potatoes into the cooker, cover the Crock-Pot Express® and cook on High Pressure for 5 more minutes. To get 5-minutes cook time, press "Steam" button and use the TIME ADJUSTMENT button to adjust the cook time to 5 minutes
5. **Pressure Release**.Release pressure, using the Natural release Method.
6. Serve and Enjoy!

Per Serving Calories: 170; Fat: 6g; Sat Fat: 2g; Carb: 2g; Fiber: 0g; Protein: 25g; Sugar: 0g; Sodium: 321 mg; Cholesterol: 70mg

Easy Pulled Pork Taco Dinner

PREP: 5 MINUTES • PRESSURE: 45 MINUTES • TOTAL: 50 MINUTES • PRESSURE LEVEL: HIGH • RELEASE: QUICK
SERVES 2

Ingredients

2 tablespoons olive oil
4 lbs. boneless pork shoulder, cut in two pieces
2 cups barbecue sauce, divided
Pinch of black pepper
Pinch of Cayenne pepper
½ cup water

Directions

1. **Preparing the Ingredients.** Preheat the Crock-Pot Express® by selecting Brown/Sauté function. Brown the pork on both sides with the oil 2-3 minutes per side. Once browned, set aside. Mix 1 cup barbecue sauce and ½ cup water into the Crock-Pot Express®. Stir to combine and add the pork back into the Crock-Pot Express® and the rest of the seasoning.
2. **High pressure for 45 minutes**. Cook on High Pressure for 45 minutes. To get 45-minutes cook time, press "Meat/Stew" button and use the TIME ADJUSTMENT button to adjust the cook time to 45 minutes.
3. **Pressure Release**. When cooking is complete, select cancel and use Natural Release (about 15 minutes).
4. **Finish the dish.** Remove the pork from the Crock-Pot Express® and shred with two forks. Strain the cooking juice through a sieve and set aside ½ cup of the juice.
5. Select Brown/sauté function, put the shredded pork back into Crock-Pot Express® \and add the 1 cup barbecue sauce and the ½ cup of cooking juice and bring it back to a simmer and mix well.
6. Put the Crock-Pot Express® on "brown/sauté" put the shredded pork back into Crock-Pot Express® and add the 1 cup barbecue sauce and the ½ cup of cooking juice and bring it back to a simmer and mix well.
7. Serve on warm taco shells with your favorite toppings or on toasted rolls.
8. For the remaining Jus you can save it and turn it into a gravy for your next meal.

Western Shoulder Ribs with Barbecue Rub and Sauce

PREP: 5 MINUTES • PRESSURE: 75 MINUTES • TOTAL: 80 MINUTES • PRESSURE LEVEL: HIGH • RELEASE: NATURAL

SERVES 6-8

Ingredients

3 pounds pork shoulder western ribs (or cut a pork shoulder into 11/2" thick strips, or pork shoulder chops)
2 teaspoons Diamond Crystal kosher salt or 1 1/2 teaspoons table salt
1 teaspoon barbecue rub
1/2 cup water
1/2 cup barbecue sauce for the cooker, plus 1/2 cup barbecue sauce to stir in at the end

Directions

1. **Preparing the Ingredients**. Sprinkle the ribs evenly with the kosher salt and the barbecue rub. Pour the 1/2 cup water into the Crock-Pot Express® , add the ribs, and then pour 1/2 cup of barbecue sauce over the ribs. (Don't stir - we want the sweet barbecue sauce to float on top to keep it from burning.)
2. **High pressure for 45 minutes**. Lock the lid on the Crock-Pot Express® and cook at high pressure for 45 minutes. To get 45-minutes cook time, press "Meat/Stew" button and use the TIME ADJUSTMENT button to adjust the cook time to 45 minutes.
3. **Pressure Release**. Let the pressure release naturally, about 15 minutes.
4. **Finish the dish**. Remove the ribs to a platter. Spoon 1/2 cup of the liquid from the pot into a measuring cup. Stir 1/2 cup of barbecue sauce into the juices in the measuring cup.
5. Serve and Enjoy!

Asian Pork

PREP: 5 MINUTES • PRESSURE: 35 MINUTES • TOTAL: 40 MINUTES • PRESSURE LEVEL: HIGH • RELEASE: NATURAL
SERVES 2

Ingredients

¼ cup hoisin sauce

2 tablespoons rice vinegar

1 tablespoon minced fresh ginger

2 teaspoons minced garlic

1 teaspoon Asian chili-garlic sauce, plus additional as desired

¾ pound pork shoulder, trimmed of as much visible fat as possible, cut into 2-inch cubes

4 slider buns or soft dinner rolls

Directions

1. **Preparing the Ingredients**. In the Crock-Pot Express® , stir together the hoisin sauce, rice vinegar, ginger, garlic, and chili-garlic sauce until thoroughly mixed. Add the pork, and toss to coat.

6. **High pressure for 35 minutes** Lock the lid in place, and bring the pot to high pressure. Cook at high pressure for 35 minutes. To get 35-minutes cook time, press "Meat/Stew" button and use the TIME ADJUSTMENT button to adjust the cook time to 35 minutes.

2. **Pressure Release**. Use the natural-release method.

3. **Finish the dish.** Unlock and remove the lid. Pour the pork and sauce through a coarse sieve; set the pork aside to cool. Return the sauce to the cooker, and let it sit for 1 to 2 minutes so any fat rises to the surface. Skim or blot off as much fat as possible and discard. Turn the Crock-Pot Express® to "brown/sauté" and simmer the sauce for about 5 minutes, or until it's the consistency of a thick tomato sauce. While the sauce thickens, shred the pork, discarding any fat or gristle. Add the shredded pork to the sauce, and heat through.

4. Serve on buns or over rice.

5. Enjoy!

PER SERVING: CALORIES: 680; FAT: 31G; SODIUM: 1,188MG; CARBOHYDRATES: 60G; FIBER: 3G; PROTEIN: 38G

Red-cooked Pork

Ingredients

2lb/ 900g fatty pork belly, cut into 1.5" or 3.8cm cubes
2 tbsp. maple syrup
3 tbsp. sherry
1 tbsp. blackstrap molasses
2 tbsp. coconut aminos
1 tsp sea salt
⅓ cup water or bone broth
1 1"/ 2.5cm length fresh ginger, peeled and smashed
A few sprigs of coriander/ cilantro leaves, to garnish

Directions

1. **Preparing the Ingredients**. Fill a pot with enough water to cover the quantity of pork cubes and bring to boil over high heat. Add the pork cubes and boil for 3 minutes, then drain and rinse off any scum or impurities. Set aside the pork cubes in a colander to drain.Heat the maple syrup in the inner pot of the Crock-Pot Express® on 'Brown/Sauté' setting.
2. Add the pork cubes to the heated maple syrup and brown the pork for approximately 10 minutes (use a splatter guard). Add the rest of the ingredients to the pot. Bring to a boil, then press cancel.
3. **High pressure for 25 minutes**. Seal the lid, and valve, then select 'Meat/Stew' setting and adjust the cooking time to 25 minutes.
4. **Pressure Release**. Allow the pressure to release naturally.
5. **Finish the dish**. Open the lid and select 'saute' setting. Bring the contents to a simmer until the sauce is sufficiently reduced and thickened to coat the pork cubes (or to your liking).
6. Serve with coriander/ cilantro leaves as a garnish. Enjoy!

Pork Chops With Applesauce

PREP: 10 MINUTES • PRESSURE: 11 MINUTES • TOTAL: 21 MINUTES • PRESSURE LEVEL: HIGH • RELEASE: NATURAL

SERVES 2-4

Ingredients

2 – 4 pork loin chops (we used center cut, bone-on)
1 tablespoon grapeseed oil or olive oil
1 small onion, sliced
3 cloves garlic, roughly minced
2 gala apples, thinly sliced
2 pieces whole cloves (optional)
1 teaspoon cinnamon powder
1 tablespoon honey
½ cup unsalted homemade chicken stock or water
2 tablespoons light soy sauce
1 tablespoon butter
Kosher salt and ground black pepper to taste
1 ½ tablespoon cornstarch mixed with 2 tablespoons water (optional)

Directions

1. **Preparing the Ingredients**. Make a few small cut around the sides of the pork chops so they will stay flat and brown evenly. Season the pork chops with generous amount of kosher salt and ground black pepper. Heat up your Crock-Pot Express®, pressing "Brown/Sauté" button. Add grapeseed oil into the pot. Add the seasoned pork chops into the pot, then let it brown for roughly 2 – 3 minutes on each side. Remove and set aside.
2. Add in the sliced onions and stir. Add a pinch of kosher salt and ground black pepper to season if you like. Cook the onions for roughly 1 minute until soften. Then, add garlic and stir for 30 seconds until fragrance.
3. Add in the thinly sliced gala apples, whole cloves (optional) and cinnamon powder, then give it a quick stir. Add the honey and partially deglaze the bottom of the pot with a wooden spoon.
4. Add chicken stock and light soy sauce, then fully deglaze the bottom of the pot with a wooden spoon. Taste the seasoning and add more salt and pepper if desired.
5. Place the pork chops back with all the meat juice into the pot.
7. **High pressure for 10 minutes**. Close lid and cook at high pressure for 10 minutes. Select 'Meat/Stew' setting and adjust the cooking time to 10 minutes.
6. **Pressure Release**. Turn off the heat and Let it fully natural release (roughly 10 minutes). Open the lid carefully.
7. **Finish the dish**. Remove the pork chops and set aside. Turn heat to medium, Press brown/sauté button. Remove the cloves and taste the seasoning one more time. Add more salt and pepper if desired. Add butter and stir until it has fully dissolved into the

sauce. Mix the cornstarch with water and mix it into the applesauce one third at a time until desired thickness.

8. Drizzle the applesauce over the pork chops and serve immediately with side dishes!
9. Enjoy!

Spare Ribs With Wine

PREP: 5 MINUTES • PRESSURE: 30 MINUTES • TOTAL: 35 MINUTES • PRESSURE LEVEL: HIGH • RELEASE: NATURAL

SERVES 2-4

Ingredients

1 pound pork spare ribs, cut into pieces
1 tablespoon oil
1 tablespoon corn starch
1 – 2 teaspoon water
Green onions as garnish
1 teaspoon fish sauce (optional)
Black Bean Marinade:
1 tablespoon black bean sauce
1 tablespoon light soy sauce
1 tablespoon Shaoxing wine
1 tablespoon ginger, grated
3 cloves garlic, minced
1 teaspoon sesame oil
1 teaspoon sugar
A pinch of white pepper

Directions

1. **Preparing the Ingredients**. Marinate the pork spare ribs with Black Bean Marinade in an oven-safe bowl. Then, sit it in the fridge for 25 minutes.
2. First, mix 1 tablespoon of oil into the marinated spare ribs. Then, add 1 tablespoon of corn starch and mix well. Finally, add 1 – 2 teaspoon of water into the spare ribs and mix well.
3. Add 1 cup of water into the Crock-Pot Express® . Place steam rack in the Crock-Pot Express® . Then, put the bowl of spare ribs on the rack.
4. **High pressure for 15 minutes**. Close lid. Pressure cook at high pressure for 15 minutes. To Get 15 Minutes, Select 'Meat/Stew' setting and adjust the cooking time to 15 minutes.
5. **Pressure Release**. Turn off the heat and Let it fully natural release (roughly 10 minutes). Open the lid carefully.
6. **Finish the dish.**Taste and add one teaspoon of fish sauce and green onions as garish if you like.
7. Serve immediately. Enjoy!

Pork Loin With Apples

PREP: 5 MINUTES • PRESSURE: 30 MINUTES • TOTAL: 35 MINUTES • PRESSURE LEVEL: HIGH • RELEASE: QUICK
SERVES 6-8

Ingredients

2 tablespoons unsalted butter
One 3-pound boneless pork loin roast
1 large red onion, halved and thinly sliced
2 medium tart green apples, such as Granny Smith, peeled, cored, and thinly sliced
4 fresh thyme sprigs
2 bay leaves
½ cup moderately sweet white wine, such as Riesling
¼ cup chicken broth
½ teaspoon salt
½ teaspoon ground black pepper

Directions

1. **Preparing the Ingredients**. Melt the butter in the Crock-Pot Express®, set on the "Browning" function. Add the pork loin and brown it on all sides, turning occasionally, about 8 minutes in all. Transfer to a large plate.
2. Add the onion to the pot; cook, stirring often, until softened, about 3 minutes. Stir in the apple, thyme, and bay leaves. Pour in the wine and scrape up any browned bits on the bottom of the pot.
3. Pour in the broth; stir in the salt and pepper. Nestle the pork loin into this apple mixture; pour any juices from the plate into the pot.
4. **High pressure for 30 minutes** Lock the lid onto the pot. Set the Crock-Pot Express® to cook at high pressure for 30 minutes. To Get 30 Minutes, Select 'Meat/Stew' setting and adjust the cooking time to 30 minutes.
5. **Pressure Release**. Use the quick-release method to bring the pot's pressure to normal.
6. **Finish the dish.** Unlock and open the cooker. Discard the bay leaves. Transfer the pork to a cutting board; let stand for 5 minutes while you dish the sauce into serving bowls or onto a serving platter. Slice the loin into ½-inch-thick rounds and lay these over the sauce.
7. Serve and Enjoy!

Tandoori BBQ Pork Ribs

PREP: 5 MINUTES • PRESSURE: 25 MINUTES • TOTAL: 30 MINUTES • PRESSURE LEVEL: HIGH • RELEASE: NATURAL

SERVES 2-4

Ingredients

2 pounds (1kilo) Pork Short-Ribs (also called Baby Back Ribs)
2 bay leaves
1" (3cm) ginger, roughly chopped
5 garlic cloves
4 tablespoons Tandoori Spice Mix (or your favorite dry rub)
3 cups water, or as needed
1½ teaspoons salt ½ cup BBQ Sauce (your favorite kind)

Directions

1. **Preparing the Ingredients**. Slice rib slabs to fit in the Crock-Pot Express® and position them in the cooker as flat as possible (this means you'll use the least amount of water- it will pressure cook faster and concentrate the flavors). Add bay leaves, ginger, garlic, salt and two tablespoons of the spice mix. Pour-in enough water to cover the meat (about 4 cups).
2. **High pressure for 22 minutes.** Close and lock the lid of the Crock-Pot Express®. Cook for 22 minutes at high pressure. To Get 22 Minutes, Select 'Meat/Stew' setting and adjust the cooking time to 22 minutes.
3. When time is up, open the Crock-Pot Express® with the Natural release method.
4. **Pressure Release**. Carefully lift the tender ribs out of the Crock-Pot Express® and lay them on a cutting board. Cover with foil and let them cool down further for another 5 minutes.
5. **Finish the dish**. Pat dry and paint on the BBQ Sauce (or spice paste). Make ahead: Stop here and wrap the meat tightly - refrigerate for up to three days. Grill, broil or barbecue for about 5 minutes per side.
6. Serve immediately. Enjoy!

Chicken Stock

PREP: 10 MINUTES • PRESSURE: 60 MINUTES • TOTAL: 70 MINUTES • PRESSURE LEVEL: HIGH • RELEASE: NATURAL

SERVES 10 Cups

Ingredients

- 2 ½ pounds chicken carcasses
- 2 onions (keep the outer layers too), diced
- 2 celery stalks, diced
- 2 carrots, diced
- 2 bay leaves
- 4 garlic cloves, crushed
- 1 teaspoon whole peppercorn
- 10 cups water
- Your favorite fresh herbs
- 1 tablespoon apple cider vinegar (optional)

Directions

1. **Preparing the Ingredients.** Optional step: Brown the chicken carcasses in your Crock-Pot Multi-Cooker® with 1 tablespoon of oil. This will slightly elevate the flavors and result in a brown stock. Then, add water to deglaze the pot with 100 ml of water.
Add all ingredients in the Crock-Pot Multi-Cooker®

2. **High pressure for 60 minutes**. Lock the lid on the Crock-Pot Multi-Cooker® and then cook for 60 minutes. To get 60-minutes cook time, press "Soup" button and use the TIME ADJUSTMEN button to adjust the cook time to 60 minutes.

3. **Pressure Release.** When the time is up, open the cooker with the Natural Release method

4. **Finish the dish** Open the lid. Strain the stock through a colander discarding the solids, and set aside to cool. Let the stock sit in the fridge until the fat rises to the top and form a layer of gel. Then, skim off the fat on the surface.
You can use the stock immediately, keep it in the fridge, or freeze it for future use.
Storage: -Silicone Mold – We love freezing our chicken stock with this mold!! After they freeze in the mold, we pop them out and store them in Ziploc freezer bags. It's a great portion for many recipes, thaws quickly, and super convenient.

Colombian Chicken Soup

PREP: 5 MINUTES • PRESSURE: HIGH • TIME UNDER PRESSURE: 17 MINUTES • RELEASE: QUICK

SERVES: 4

Ingredients

1 medium yellow onion, cut in half

2 medium carrots, cut in half crosswise

2 ribs celery, cut in half crosswise

3 bone-in chicken breasts (about 2 pounds, or 907 g)

5 cups (1.2 L) water

1 1/2 teaspoons kosher salt

1 1/2 pounds (680 g) Yukon gold potatoes, cut into 1/2-inch (13 mm) pieces

1 ear corn, cut into 4 pieces

1/4 teaspoon freshly ground black pepper

1 avocado

1/4 cup (60 g) sour cream

1 tablespoon (9 g) capers, rinsed

1 teaspoon dried oregano

8 sprigs fresh cilantro

1 lime, quartered

Directions

1. **Preparing the Ingredients** To the Crock-Pot Multi-Cooker ®, add the onion, carrots, celery, chicken, water, and salt.
2. **High pressure for 15 minutes.** Lock the lid on the Crock-Pot Multi-Cooker® and then cook for 15 minutes. To get 15-minutes cook time, press "Poultry" button.
3. **Pressure Release.** Use the "Quick Release" method to vent the steam, then open the lid. Transfer the chicken to a large bowl. When cool enough to handle, shred into pieces, discarding the skin and bones.
   Discard the onion, carrots, and celery. Add the potatoes and corn to the broth.
4. **High pressure for 2 minutes.** Lock the lid on the Crock-Pot Multi-Cooker® and then cook for 2 minutes. To get 2-minutes cook time, press "Steam" button and use the TIME ADJUSTMENT button to adjust the cook time to 2 minutes.
5. **Pressure Release.** Use the "Quick Release" method to vent the steam, then open the lid.
6. **Finish the dish.** Stir in the chicken and pepper.
   Divide the soup among bowls. Peel, pit, and slice the avocado. Top the soup with the avocado, sour cream, capers, oregano, and cilantro.
   Serve with the lime quarters for squeezing.
   Enjoy!

Corn Chowder

PREP: 5 MINUTES • PRESSURE: HIGH • TIME UNDER PRESSURE: 6 MINUTES • RELEASE: NATURAL

SERVES: 4-6

Ingredients

- 1 tablespoon olive oil
- 1 medium yellow onion, diced
- 1 medium red bell pepper, stemmed, seeded, and diced
- 1 medium green bell pepper, stemmed, seeded, and diced
- 4 cups salt-free Vegetable Stock
- 3 small unpeeled red potatoes, cubed
- 4 cups fresh or thawed frozen corn kernels
- 3 tablespoons unsalted butter
- 3 tablespoons all-purpose flour
- 1 cup whole milk
- 3 teaspoons salt
- ½ teaspoon freshly ground black pepper, plus more if desired
- 6 slices crisp-cooked prosciutto, for serving

Directions

1. **Preparing the Ingredients** Heat the Crock-Pot Multi-Cooker ® using the "Sauté" function, add the oil, onion, and red and green bell peppers and sauté, stirring infrequently, until the onion is translucent, about 5 minutes. Stir in the stock, potatoes, and corn.

2. **High pressure for 6 minutes.** Lock the lid on the Crock-Pot Multi-Cooker® and then cook for 6 minutes. To get 6-minutes cook time, press "Rice/Risotto" button. Meanwhile, make a blond roux. In a small saucepan over low heat, mix together the butter and flour and cook, stirring constantly, until the butter has melted and the mixture foams and forms a thick paste. Remove from the heat.

3. **Pressure Release.** When the time is up, open the cooker with the Normal Release method.

4. **Finish the dish.** Stir the roux, milk, salt, and black pepper into the chowder. Return the uncovered cooker using the "Sauté, and simmer the soup, stirring occasionally, until it reaches the desired thickness.
   Serve with a crispy slice of prosciutto in each bowl.

5. Enjoy!

Creole White Bean Soup

PREP: 5 MINUTES • PRESSURE: HIGH • TIME UNDER PRESSURE: 12 MINUTES • RELEASE: QUICK

SERVES: 2-4

Ingredients

1 tablespoon kosher salt

1 quart water

6 ounces dried navy beans

1 tablespoon olive oil

1½ ounces ham, diced (about ⅓ cup)

⅓ cup chopped onion

1 tablespoon minced garlic (about 3 medium cloves)

1 teaspoon Creole or Cajun seasoning

¼ teaspoon ground cayenne pepper (optional)

3 cups Chicken Stock or low-sodium broth, plus additional as needed

2 tablespoons Creole or other whole-grain mustard

½ teaspoon hot pepper sauce (such as Tabasco or Crystal), plus additional as needed

1 teaspoon Worcestershire sauce

1 small tomato, seeded and diced, or ⅓ cup canned diced tomatoes

¼ cup chopped scallions

3 cups loosely packed arugula

Directions

1. **Preparing the Ingredients.** In a large bowl, dissolve the kosher salt in the water. Add the beans, and soak at room temperature for 8 to 24 hours. Drain and rinse.

   Set the Crock-Pot Multi-Cooker® to "brown," heat the olive oil until it shimmers and flows like water. Add the ham, and cook for 2 to 3 minutes, or until it just starts to brown. Add the onion and garlic, and cook for about 2 minutes, or until the onion pieces start to separate and the garlic becomes fragrant. Stir in the Creole seasoning and cayenne pepper (if using), and cook for 1 minute, stirring to coat the ham and vegetables.

   Add the Chicken Stock; then pour in the beans.

2. **High pressure for 12 minutes.** Lock the lid on the Crock-Pot Multi-Cooker® and then cook for 12 minutes. To get 12-minutes cook time, press "Soup" button and use the TIME ADJUSTMENT button to adjust the cook time to 12 minutes.

3. **Pressure Release.** Use the quick-release method.

4. **Finish the dish.** Unlock and remove the lid. Turn the Crock-Pot Multi-Cooker ® to "brown, stir in the mustard, hot pepper sauce, and Worcestershire sauce, and simmer for 3 minutes. Taste and adjust the seasoning, adding more hot sauce or Creole seasoning if you want it spicier. If the soup is too spicy or too thick, add more stock. Add the tomato, scallions, and arugula, and simmer for about 4 minutes, or until the arugula is wilted and the tomatoes are heated through. Ladle into bowls, and serve.

PER SERVING (MAIN COURSE): CALORIES: 497; FAT: 14G; SODIUM: 734MG; CARBOHYDRATES: 66G; FIBER: 25G; PROTEIN: 30G

Spicy Chicken and Tomato Soup

PREP: 10 MINUTES • PRESSURE: 12 MINUTES • TOTAL: 20 MINUTES • PRESSURE LEVEL: LOW • RELEASE: NATURAL

SERVES 4

Ingredients

- 1/2 teaspoon ground cumin
- 1 (15.5-ounce) *can navy beans, rinsed and drained
- 1 (14.5-ounce) *can no-salt-added stewed tomatoes
- 1 (14-ounce) *can fat-free, less-sodium chicken broth
- 1 chipotle chile, canned in adobo sauce, finely chopped
- 2 cups chopped cooked chicken breast (about 1/2 pound)
- 1 tablespoon extra virgin olive oil
- 1/2 cup reduced-fat sour cream
- 1/4 cup chopped fresh cilantro

Directions

1. **Preparing the Ingredients.** Select "Saute" programme to heat oil. Sauté the garlic, onion, carrot and celery. Pour in the diced tomatoes with juice.
   Add the bacon, chicken, rosemary and bay leaf. Stir to combine.
   Pour in the chicken stock. Add the pasta.
2. **High pressure for 12 minutes.** Lock the lid on the Crock-Pot Multi-Cooker® and then cook for 12 minutes. To get 12-minutes cook time, press Soup button and use the TIME ADJUSTMENT button to adjust the cook time to 12 minutes.
3. **Pressure Release.** Run quick release.
4. **Finish the dish.** Drain out the chicken. Shred with two forks. Toss back into the soup. Taste first. Season with salt if necessary. Sprinkle pepper. Garnish with chopped parsley. Serve immediately with crusty bread if desired.

Chicken Stew

PREP: 8 MINUTES • PRESSURE: 30 MINUTES • TOTAL: 38 MINUTES • PRESSURE LEVEL: HIGH • RELEASE: NATURAL
SERVES 6-8

Ingredients

- 1 teaspoon vegetable oil
- 6 chicken thighs (about 3 1/2 pounds) or a cut-up whole chicken
- 2 teaspoons kosher salt
- 1 large onion, diced
- 1 stalk celery, diced
- 1/4 pound baby carrots, cut into 1/2 inch slices
- 2 tablespoons tomato paste
- 1/2 teaspoon dried thyme
- 1/2 teaspoon Diamond Crystal kosher salt
- 1/2 cup white wine
- 2 cups chicken stock, preferably homemade
- 15-ounce can diced tomatoes
- 3/4 pound baby carrots
- 1 1/2 pounds new potatoes

Directions

1. **Preparing the Ingredients.** Season the chicken with 2 teaspoons salt. Select the "Browning" Mode. Add 1 teaspoon of vegetable oil, wait until shimmering. Brown the chicken in 2 batches, three pieces in each batch. Sear the chicken for 4 minutes per side, or until well browned.

   Once all the chicken is browned, pour off all but 1 tablespoon of the fat in the cooker.

   Sauté the aromatics: Add the onion, celery, sliced carrots, tomato paste, and thyme to the pot. Sprinkle with 1/2 teaspoon salt. Sauté for five minutes, or until the onions are softened. Add the white wine to the pot, bring to a simmer, and scrape the bottom of the pot to loosen any browned bits. Simmer the wine until reduced by half - about 3 minutes.

   Stir in the chicken stock, then add the chicken thighs and any chicken juices from the bowl. Pour the tomatoes on top, but don't stir. Put a steamer basket on top of everything in the pot - don't worry if it's a little submerged, it will be fine - and put the potatoes and carrots in the steamer basket.

2. **High pressure for 30 minutes.** Lock the lid on the Crock-Pot Multi-Cooker® and then cook for 30 minutes. To get 30-minutes cook time, press "Soup" button.

3. **Pressure Release.** Use natural-release method for 15 minutes, then quick release.

4. **Finish the dish.** Carefully lift the steamer basket of potatoes and carrots out of the pot, then scoop the chicken pieces out with a slotted spoon. Cut the potatoes in half, and then stir the carrots and potatoes back into the stew. Shred the chicken, discarding the skin, bones, and gristle, and stir the shredded chicken meat back into the stew. Taste for seasoning, adding salt and pepper if necessary.

5. Serve and Enjoy.

Steamed Mussels in Porter Cream Sauce

PREP: 5 MINUTES • PRESSURE: 1 MINUTES • TOTAL: 6 MINUTES • PRESSURE LEVEL: HIGH • RELEASE: QUICK

SERVES 2-4

Ingredients
   tablespoon olive oil
   2 garlic cloves, minced
   2 scallions, minced (about ⅓ cup)
   1 (12-ounce) bottle porter or other dark beer
   ⅛ teaspoon red pepper flakes
   2 pounds mussels, scrubbed and debearded
   2 tablespoons heavy (whipping) cream

Directions
1. **Preparing the Ingredients.** Set the Crock-Pot Multi-Cooker® to "browning," heat the olive oil until it shimmers and flows like water. Add the garlic and scallions, and cook for about 3 minutes, stirring, until the scallions just begin to brown. Pour in the beer, stirring for 1 minute, or until the foam dissipates. Add the red pepper flakes and mussels, and stir to coat with the liquid.
2. **High pressure for 1 minute.** Bring the cooker to high pressure by pressing the "Steam" button. Allow to cook for 1 minute and press Stop.
3. **Pressure Release** Use the quick-release method.
4. **Finish the dish** Unlock and remove the lid. The mussels should be opened; if not, replace but don't lock the lid, turn the Crock-Pot Multi-Cooker® to "brown", for 1 to 2 minutes more. Discard any mussels that still have not opened. Stir in the heavy cream; then pour the mussels with their sauce into a large serving bowl, and enjoy.

PER SERVING (MAIN COURSE): CALORIES: 584; FAT: 23G; SODIUM: 1,313MG; CARBOHYDRATES: 25G; FIBER:
0G; PROTEIN: 56G

Lemon and Dill Fish Packets

PREP: 10 MINUTES • PRESSURE: 5 MINUTES • TOTAL: 15 MINUTES • PRESSURE LEVEL: HIG • RELEASE: QUICK

SERVES 2

Ingredients

2 tilapia or cod fillets
Salt, pepper, and garlic powder
2 sprigs fresh dill
4 slices lemon
2 tablespoons butter

Directions

1. **Preparing the Ingredients.** Lay out 2 large squares of parchment paper.

   Place a fillet in the center of each parchment square, and then season with a generous amount of salt, pepper, and garlic powder.

   On each fillet, place in order: 1 sprig of dill, 2 lemon slices, and 1 tablespoon of butter.

   For best results, place a small metal rack or trivet at the bottom of your Crock-Pot Multi-Cooker®.

   Pour 1 cup of water into the cooker to create a water bath.

   Close up parchment paper around the fillets, folding to seal, and then place both packets on metal rack inside cooker.

2. **High pressure for 5 minutes**. Lock the lid on the Crock-Pot Multi-Cooker® and then cook for 5 minutes. To get 5-minutes cook time, press "Soup" button, allow to cook for 5 minute and press Stop.

3. **Pressure Release** Perform a quick release to release the cooker's pressure. Unwrap packets and serve.

   There is no need to remove the fish from the packets before serving. In fact, it makes a really nice presentation.

Red Curry Cod With Red Beans

PREP: 5 MINUTES • PRESSURE: 5 MINUTES • TOTAL: 10 MINUTES • PRESSURE LEVEL: HIHG • RELEASE: QUICK

SERVES 4

Ingredients

- 1 can (13 1/2 ounces, or 400 ml) unsweetened coconut milk
- 2 tablespoons (30 g) red Thai curry paste
- 1 tablespoon (8 g) finely grated fresh ginger
- 1 1/2 pounds (680 g) cod or halibut fillet, cut into 2-inch (5 cm) pieces
- 8 ounces (225 g) green beans
- 1/2 cup (8 g) fresh cilantro leaves
- 2 scallions (white and light green parts), thinly sliced
- 1 lime, quartered

Directions

1. **Preparing the Ingredients** To the Crock-Pot Multi-Cooker®, add the coconut milk, curry paste, and ginger, and whisk together.
   Add the cod. Lay the green beans on top.
2. **High pressure for 5 minutes.** Lock the lid on the Crock-Pot Multi-Cooker® and then cook for 5 minutes. To get 5-minutes cook time, press Soup button and use the TIME ADJUSTMENT button to adjust the cook time to 5 minutes.
3. **Pressure Release.** Use the "Quick Release" method to vent the steam, then open the lid.
4. Top the curry with the cilantro and scallions, and serve with the lime quarters for squeezing.

Smoked Salmon Chowder

PREP: 5 MINUTES • PRESSURE: 6 MINUTES • TOTAL: 10 MINUTES • PRESSURE LEVEL: LOW • RELEASE: QUICK
SERVES 6

Ingredients

1 tablespoon unsalted butter
2 large scallions, chopped
½ teaspoon kosher salt, plus additional for seasoning
1 tablespoon all-purpose flour
¼ cup dry white wine, dry vermouth, or dry sherry
2½ cups whole milk
2 small (or 1 medium) red or Yukon gold potatoes (about 5 ounces), peeled and cut into ½-inch cubes
1 (4- or 5-ounce) salmon fillet, skinned
1½ ounces hot-smoked salmon, chopped or flaked into small chunks
3 teaspoons chopped fresh dill, divided
1 teaspoon lemon zest
Freshly ground black pepper

Directions

1. **Preparing the Ingredients.** The Crock-Pot Multi-Cooker® set to "browning," melt the butter. When the butter is foaming, add the scallions and sprinkle with ½ teaspoon of kosher salt. Cook for 1 minute, stirring, until softened. Add the flour, and cook for 2 to 3 minutes, or until it turns a very light tan color. Add the white wine, and cook for about 2 minutes, or until the mixture has thickened. Add the milk, and whisk until the mixture is smooth.
   Add the potatoes.

2. **High pressure for 5 minutes.** Lock the lid on the Crock-Pot Multi-Cooker® and then cook for 5 minutes. To get 5-minutes cook time, press "Soup" button and use the TIME ADJUSTMENT button to adjust the cook time to 5 minutes

3. **Pressure Release.** Use the quick-release method.
   Unlock and remove the lid. Add the raw salmon fillet, and replace the lid.

4. **High pressure for 1 minutes** Lock the lid in place, bring the cooker to high pressure by pressing the STEAM button. Allow to cook for 1 minute and press STOP.

5. **Pressure Release.** After cooking, use the natural method to release pressure for 4 minutes, then the quick method to release the remaining pressure.

6. **Finish the dish** Unlock and remove the lid. Using a large slotted spoon or fish spatula, remove the salmon fillet to a plate or cutting board. Use a fork to break it into chunks. Don't worry if the fish is not completely cooked; it will finish cooking later.
   Turn the Crock-Pot Multi-Cooker® to "brown, add the salmon chunks, smoked salmon, 2 teaspoons of dill, and the lemon zest, an sim for 1 to 2 minutes, or until the fish is heated through. Adjust the seasoning with additional kosher salt and pepper. Sprinkle the remaining 1 teaspoon of dill over the soup just before serving.

Fish *and* Vegetable "Tagine" *with* Chermoula

PREP: 5 MINUTES • PRESSURE: 5 MINUTES • TOTAL: 10 MINUTES • PRESSURE LEVEL: HIGH • RELEASE: QUICK

SERVES 2

Ingredients

- For The Chermoula
  4 large garlic cloves
  1 cup fresh cilantro leaves
  1 cup fresh parsley leaves
  ¼ cup freshly squeezed lemon juice (about 2 lemons)
  1½ teaspoons kosher salt
  1 heaping teaspoon ground sweet paprika
  ¼ teaspoon ground cumin
  ¼ teaspoon ground cayenne pepper
  2 tablespoons olive oil
- For The Fish And Vegetables
  2 (7-ounce) tilapia fillets
  ¼ teaspoon kosher salt
  10 ounces Yukon gold potatoes (about 2 medium or 3 small), peeled and sliced ¼ inch thick
  ½ medium red bell pepper, cut into bite-size chunks
  ½ medium green bell pepper, cut into bite-size chunks
  1 very small onion, sliced
  ¼ cup water
  1 large tomato, seeded and diced

To make the chermoula

1. **Preparing the Ingredients.** Into the chute of a small running food processor, drop the garlic cloves, one at a time, and process until minced. Add the cilantro, parsley, lemon juice, kosher salt, paprika, cumin, and cayenne pepper, and process until mostly smooth. With the processor still running, slowly drizzle in the olive oil, and process until the sauce is emulsified.

   If you don't have a food processor, finely mince the garlic, cilantro, and parsley. Transfer to a small bowl, and stir in the lemon juice, kosher salt, paprika, cumin, and cayenne pepper. Slowly whisk in the olive oil. The sauce won't be as smooth as if prepared in a food processor, but it will taste good.

To make the fish and vegetables

1. **Preparing the Ingredients.** Sprinkle both sides of the fish fillets lightly with the kosher salt, and brush with 3 tablespoons of chermoula. Refrigerate the fish.

   Add the potato slices, red bell pepper, green bell pepper, and onion. Pour in ⅓ cup of chermoula, and gently toss the vegetables to coat. Pour the water over the vegetables.

2. **High pressure for 5 minutes.** Lock the lid on the Crock-Pot Multi-Cooker® and then cook for 5 minutes. To get 5-minutes cook time, press "Soup" button and use the TIME ADJUSTMENT button to adjust the cook time to 5 minutes

3. **Pressure Release** Use the quick-release method.

   Unlock and remove the lid. Sprinkle the tomato over the vegetables in the Crock-Pot Multi-Cooker®, and lay the fillets on top. Drizzle with the remaining chermoula.

4. **High pressure for 1 minute.** Lock the lid in place again; bring the cooker to high pressure by pressing the "STEAM" button. Allow to cook for 1 minute and press STOP When the timer goes off, turn the cooker off. ("Warm" setting, turn off).

5. **Pressure Release** After cooking, use the natural method to release pressure for 4 minutes, then the quick method to release the remaining pressure.

6. **Finish the dish** Unlock and remove the lid. Using a large spatula, carefully remove the fish fillets and vegetables and divide them between 2 plates. Spoon any residual sauce over the fish, and serve.

PER SERVING: CALORIES: 479; FAT: 17G; SODIUM: 2,157MG; CARBOHYDRATES: 43G; FIBER: 7G; PROTEIN: 44G

Coconut Fish Curry

PREP: 5 MINUTES • PRESSURE: 15 MINUTES • TOTAL: 20 MINUTES • PRESSURE LEVEL: HIGH • RELEASE: QUICK
SERVES 2-4

Ingredients

1-1.5 lb. (500-750g) Fish steaks or fillets, rinsed and cut into bite-size pieces (fresh or frozen and thawed)

1 Tomato, chopped (or a heaping cup of cherry tomatoes)

2 Green Chiles, sliced into strips

2 Medium onions, sliced into strips

2 Garlic cloves, squeezed

1 Tbsp. freshly grated Ginger, or $\frac{1}{8}$ tsp. Ginger Powder

6 Curry leaves, or Bay Laurel Leaves, or Kaffir Lime Leaves, or Basil

1 Tbsp. ground Coriander

2 tsp. ground Cumin

$\frac{1}{2}$ tsp. ground Turmeric

1 tsp. Chili Powder , or 1 tsp. of Hot Pepper Flakes

$\frac{1}{2}$ tsp. Ground Fenugreek (Methi)

3 Tbsp. of Curry Powder mix. (instead of the 5 spices noted above)

2 cups or (500ml) un-sweetened Coconut Milk

Salt to taste (I used about 2 tsp.)

Lemon juice to taste (I used the juice from $\frac{1}{2}$ lemon)

Directions

1. **Preparing the Ingredients.** In the preheated Crock-Pot Multi-Cooker®, add a swirl of oil and then drop in the curry leaves and lightly fry them until golden around the edges (about 1 minute).

   Then add the onion, garlic and ginger and Sauté until the onion is soft.

   Add all of the ground spices: Coriander, Cumin, Turmeric, Chili Powder and Fenugreek and Sauté them together with the onions until they have released their aroma (about 2 minutes).

   De-glaze with the coconut milk making sure to un-stick anything from the bottom of the cooker and incorporate it in the sauce.

   Add the Green Chiles, Tomatoes and fish pieces. Stir to coat the fish well with the mixture.

2. **High pressure for 3 minutes.** Lock the lid on the Crock-Pot Multi-Cooker® and then cook for 3 minutes. To get 3-minutes cook time, press "Steam" button and use the TIME ADJUSTMENT button to adjust the cook time to 3 minutes.

3. **Pressure Release**. Use the "Quick Release" method to vent the steam, then open the lid.

4. **Finish the dish.** Add salt to taste and spritz with lemon juice just before serving.

   Serve alone, or with steamed rice.

Collard Greens In A Tomato Sauce

PREP: 5 MINUTES • PRESSURE: 6 MINUTES • TOTAL: 11 MINUTES • PRESSURE LEVEL: HIGH • RELEASE: QUICK
SERVES 6

Ingredients

    2 tablespoons olive oil
    1 tablespoon minced garlic
    ½ teaspoon red pepper flakes
    1½ pounds collard greens, tough stems removed, the leaves chopped (about 8 packed cups)
    ½ cup canned tomato puree
    ½ cup vegetable or chicken broth
    ½ cup moderately sweet white wine, such as a dry Riesling
    ½ teaspoon salt

Directions

1. **Preparing the Ingredients.** Heat the oil in the Crock-Pot Multi-Cooker® turned to the "Browning" function. Add the garlic and red pepper flakes; cook, stirring all the while, until aromatic, less than 1 minute.

   Add the collards; toss over the heat for 2 minutes. Add the tomato puree, broth, wine, and salt, and stir well.

2. **High pressure for 6 minutes.** Lock the lid on the Crock-Pot Multi-Cooker® and then cook for 6 minutes. To get 6-minutes cook time, press "Rice/Risotto" button and use the TIME ADJUSTMENT button to adjust the cook time to 6 minutes.

3. **Pressure Release.** Use the quick-release method.

   Unlock and open the cooker. Stir well before serving.

SMASHED SWEET POTATOES WITH PINEAPPLE AND GINGER

PREP: 5 MINUTES • PRESSURE: 12 MINUTES • TOTAL: 17 MINUTES • PRESSURE LEVEL: HIGH • RELEASE: QUICK

SERVES 8

Ingredients

4 pounds medium sweet potatoes (about 6 potatoes), peeled and cut into 1½-inch chunks

3 tablespoons unsalted butter

½ teaspoon ground ginger

¼ teaspoon ground cinnamon

¼ teaspoon grated nutmeg

2 tablespoons frozen unsweetened pineapple juice concentrate, thawed

1 teaspoon salt

Directions

1. **Preparing the Ingredients.** Place the sweet potatoes and 3 cups water in the Crock-Pot Multi-Cooker®

2. **High pressure for 12 minutes.** Lock the lid on the Crock-Pot Multi-Cooker® and then cook for 12 minutes. To get 12-minutes cook time, press "Soup" button, and use the TIME ADJUSTMENT button to adjust the cook time to 12 minutes.

3. **Pressure Release** Use the quick-release method to bring the pot's pressure back to normal.

4. **Finish the dish.** Unlock and open the cooker. Drain the sweet potatoes in a colander set in the sink. Turn the electric cooker to its browning function. Melt the butter in the cooker, then add the ginger, cinnamon, and nutmeg; cook until aromatic, stirring constantly, less than 1 minute.

   Stir in the pineapple juice concentrate and salt, and then turn off the electric cooker. Add the potatoes and stir well with a wooden spoon, smashing them a bit, until you have a vaguely smooth puree with chunks of sweet potato.

Roasted Rainbow Fingerling Potatoes

PREP: 5 MINUTES • PRESSURE: 20 MINUTES • TOTAL: 25 MINUTES • PRESSURE LEVEL: HIGH • RELEASE: QUICK

SERVES 4

Ingredients

- ½ cup (100 g) diced onion
- 2 tbsp. (30 ml) ghee
- 1 tbsp. (15 ml) olive oil
- 2 lb. (907 g) rainbow fingerling potatoes
- Up to 1 tsp (5 g) sea salt
- ¼ tsp black pepper
- ½ tsp onion powder
- ½ tsp paprika

Directions

1. **Preparing the Ingredients.** Begin by sautéing the onion in your Crock-Pot Multi-Cooker® in the ghee and olive oil for 5 minutes.

   Add in the potatoes and seasonings and secure the lid.

2. **High pressure for 20 minutes.** Lock the lid on the Crock-Pot Multi-Cooker® and then cook for 20 minutes. To get 20-minutes cook time, press "Meat/Stew" button and use the TIME ADJUSTMENT button to adjust the cook time to 20 minutes.

3. **Pressure Release** Quick-release the pressure valve when complete and carefully remove the lid.

4. Serve warm.

Potato Salad

PREP: 5 MINUTES • PRESSURE: 10 MINUTES • TOTAL: 15 MINUTES • PRESSURE LEVEL: HIGH • RELEASE: QUICK

SERVES 4

Ingredients

24 oz. (680 g) Yukon gold potatoes, peeled and diced
½ cup (118 ml) water
¼ cup + 1 tbsp. (75 ml) high-quality store-bought mayonnaise
1 tbsp. (15 ml) organic yellow mustard
½ sweet onion, minced
1 rib celery, minced
½ tsp celery salt
¼ tsp dried or fresh dill, minced
1 tsp (5 ml) apple cider vinegar
Pinch ground black pepper
Optional: paprika to garnish

Directions

1. **Preparing the Ingredients.** Place the diced potatoes and water into the bowl of your Crock-Pot Multi-Cooker®.
2. **High pressure for 10 minutes.** Lock the lid on the Crock-Pot Multi-Cooker® and then cook for 10 minutes. To get 10-minutes cook time, press "Steam" button.
   Allow the potatoes to cook.
3. **Pressure Release** Quick-release the pressure valve once the cycle is complete.
4. **Finish the dish** Remove the lid once safe to do so. Drain the water from the potatoes (unless you want soupy potato salad) and stir in all the remaining ingredients.
5. Sprinkle a bit of paprika on top to garnish.
   Serve.

One-Pot Pasta Puttanesca

PREP: 5 MINUTES • PRESSURE: 8 MINUTES • TOTAL: 13 MINUTES • PRESSURE LEVEL: HIGH • RELEASE: QUICK
SERVES 4

Ingredients

- 2 tablespoons olive oil
- 1 small red onion, chopped
- 1 tablespoon drained and rinsed capers, minced
- 1 tablespoon minced garlic
- 1 pound eggplant (about 1 large), stemmed and diced (no need to peel)
- 2 medium yellow bell peppers, stemmed, cored, and chopped
- One 28-ounce can diced tomatoes (about 3½ cups)
- 1¼ cups vegetable broth
- 2 tablespoons canned tomato paste
- 2 teaspoons dried rosemary
- 1 teaspoon dried thyme
- ½ teaspoon ground black pepper
- 8 ounces dried whole wheat ziti

Directions

1. **Preparing the Ingredients.** Heat the oil in the Crock-Pot Multi-Cooker® turned to the "Browning" function. Add the onion, capers, and garlic; cook, stirring often, just until the onion first begins to soften, about 2 minutes.

   Add the eggplant and bell peppers; cook, stirring often, for 1 minute. Mix in the tomatoes, broth, tomato paste, rosemary, thyme, and pepper, stirring until the tomato paste coats everything. Stir in the ziti until coated.

2. **High pressure for 8 minutes.** Lock the lid on the Crock-Pot Multi-Cooker® and then cook for 8 minutes. To get 8-minutes cook time, press "Rice/Risotto" button and use the TIME ADJUSTMENT button to adjust the cook time to 8 minutes.

3. **Pressure Release.** Use the quick-release method to drop the pressure in the pot back to normal.

4. Unlock and open the cooker. Stir well before serving.

Ratatouille

PREP: 5 MINUTES • PRESSURE: 4 MINUTES • TOTAL: 9 MINUTES • PRESSURE LEVEL: HIGH • RELEASE: QUICK

SERVES 4

Ingredients

Kosher salt, for salting and seasoning

1 small eggplant, peeled and sliced ½ inch thick

1 medium zucchini, sliced ½ inch thick

2 tablespoons olive oil

1 cup chopped onion

3 garlic cloves, minced or pressed

1 small green bell pepper, cut into ½-inch chunks (about 1 cup)

1 small red bell pepper, cut into ½-inch chunks (about 1 cup)

1 rib celery, sliced (about 1 cup)

1 (14.5-ounce) can diced tomatoes, undrained

¼ cup water

½ teaspoon dried oregano

¼ teaspoon freshly ground black pepper

2 tablespoons minced fresh basil

¼ cup pitted green or black olives (optional)

Directions

1. **Preparing the Ingredients.** Place a rack over a baking sheet. With kosher salt, very liberally salt one side of the eggplant and zucchini slices, and place them, salted-side down, on the rack. Salt the other side. Let the slices sit for 15 to 20 minutes, or until they start to exude water (you'll see it beading up on the surface of the slices and dripping into the sheet pan). Rinse the slices, and blot them dry. Cut the zucchini slices into quarters and the eggplant slices into eighths.

   Turn the Crock-Pot Multi-Cooker® to "brown," heat the olive oil until it shimmers and flows like water. Add the onion and garlic, and sprinkle with a pinch or two of kosher salt. Cook for about 3 minutes, stirring, until the onions just begin to brown.

   Add the eggplant, zucchini, green bell pepper, red bell pepper, celery, and tomatoes with their juice, water, and oregano.

2. **High pressure for 4 minutes.** Lock the lid on the Crock-Pot Multi-Cooker® and then cook for 4 minutes. To get 4-minutes cook time, press "Steam" button and use the TIME ADJUSTMENT button to adjust the cook time to 4 minutes.

3. **Pressure Release.** Use the quick-release method.

4. **Finish the dish.** Unlock and remove the lid. Stir in the pepper, basil, and olives (if using). Taste, adjust the seasoning as needed, and serve.

   While this vegetable dish is usually served on its own, it's great tossed with cooked pasta or served over polenta.

PER SERVING: CALORIES: 149; FAT: 8G; SODIUM: 55MG; CARBOHYDRATES: 20G; FIBER: 8G; PROTEIN: 4G

Rice And Kale

PREP: 5 MINUTES • PRESSURE: 15 MINUTES • TOTAL: 20 MINUTES • PRESSURE LEVEL: HIGH • RELEASE: QUICK
SERVES 6

Ingredients

    1 tablespoon olive oil
    1 tablespoon minced garlic
    ½ teaspoon cumin seeds
    8 ounces kale, washed, stemmed, and chopped (about 2 cups packed)
    3 cups vegetable, chicken, or beef broth
    1⅓ cups long-grain white basmati rice

Directions

1. **Preparing the Ingredients**. Heat the oil in the Crock-Pot Express® turned to the "browning" function. Add the garlic and cumin; stir until aromatic, until the cumin seeds start to pop, about 1 minute.
2. Add the kale and stir until wilted, about 1 minute. Stir in the broth and scrape up any browned bits in the bottom of the cooker. Add the rice and stir well.
3. **High pressure for 15 minutes**. Lock the lid on the Crock-Pot Multi-Cooker® and then cook for 15 minutes. To get 15-minutes cook time, press "Rice/Risotto" button and use the TIME ADJUSTMENT button to adjust the cook time to 15 minutes.
4. **Pressure Release**. Bring the pot's pressure back to normal with the quick-release method but do not open the cooker. Set aside for 10 minutes to steam the rice.
5. **Finish the dish.** Unlock and open the cooker. Stir before serving.

Arroz Verde

PREP: 5 MINUTES • PRESSURE: 8 MINUTES • TOTAL: 13 MINUTES • PRESSURE LEVEL: HIGH • RELEASE: QUICK

SERVES 4- 6

Ingredients

1 medium poblano chile, stemmed, cored, seeded, and cut into 2 or 3 strips

1 small jalapeño pepper, stemmed, cored, seeded, and cut into 2 or 3 strips (optional)

1 tablespoon olive oil

⅓ cup chopped onion

½ teaspoon kosher salt, plus additional for seasoning

¾ cup long-grain white rice

1 cup plus 2 tablespoons Chicken Stock or low-sodium broth

12 tablespoons minced fresh cilantro, divided

Directions

1. **Preparing the Ingredients**
2. **High pressure for 50 minutes**.
3. **Pressure Release**
4. **Finish the dish.**

1. **Preparing the Ingredients**. Place the poblano and jalapeño pepper (if using) pieces skin-side up on an aluminum foil–lined baking sheet. Preheat the broiler. Place the baking sheet close to the broiler element. Cook for 3 to 10 minutes, depending on the strength of the broiler element, or until the skins are blackened. Remove from the broiler, carefully pull the foil up around the pepper strips, and fold over to seal. Let sit for 5 to 10 minutes, or until cool enough to handle. Peel the skin off the strips and discard. Dice the peppers.
2. Turn the Crock-Pot Express® to "brown/sauté" heat the olive oil until it shimmers and flows like water. Add the onion, and sprinkle with a pinch or two of kosher salt. Cook for about 3 minutes, stirring, until the onions just begin to brown. Add the rice, and stir to coat with the olive oil. Pour in the Chicken Stock, and add the ½ teaspoon of kosher salt, the chopped chiles, and about 6 tablespoons of cilantro. Bring to a simmer.
3. **High pressure for 8 minutes**. Lock the lid in place, and bring the pot to high pressure. Cook at high pressure for 8 minutes. To get 8-minutes cook time, press "Rice/Risotto" button and use the TIME ADJUSTMENT button to adjust the cook time to 8 minutes.
4. **Pressure Release**. Use the quick-release method.
5. **Finish the dish.** Open the pot, and gently stir in the remaining 6 tablespoons of cilantro. Replace the lid, but *don't lock* it. Let the rice steam for 4 minutes more. Fluff gently with a fork before serving. To save time, use a small can of chopped green chiles rather than roasting your own. They won't have the complex flavor of the poblanos, but they make this dish much easier. Rinse and drain them, then add them to the rice at step 3.

PER SERVING: CALORIES: 168; FAT: 4G; SODIUM: 313MG; CARBOHYDRATES: 30G; FIBER: 1G; PROTEIN: 3G

Rice And Mushrooms

PREP: 5 MINUTES • PRESSURE: 15 MINUTES • TOTAL: 20 MINUTES • PRESSURE LEVEL: HIGH • RELEASE: QUICK
SERVES 6

Ingredients

2 tablespoons peanut oil

8 medium scallions, thinly sliced

8 ounces baby bella or cremini mushrooms, thinly sliced

1½ cups long-grain white rice, preferably jasmine

1 tablespoon minced fresh ginger

3 cups chicken broth

2 tablespoons soy sauce

2 tablespoons mirin

Directions

1. **Preparing the Ingredients**. Heat the oil in the Crock-Pot Express® turned to the "browning" function. Add the scallions and mushrooms; cook, stirring often, until both soften and the mushrooms give off their liquid, about 5 minutes. Add the rice and ginger; stir for 1 minute. Pour in the broth, soy sauce, and mirin; scrape up any browned bits in the bottom of the cooker.

2. **High pressure for 15 minutes.** Lock the lid onto the pot. Cook at high pressure for 15 minutes. To get 15-minutes cook time, press "Rice/Risotto" button and use the TIME ADJUSTMENT button to adjust the cook time to 15 minutes.

3. **Pressure Release**. Use the quick-release method to return the pot's pressure to normal but do not open the cooker. Set aside for 10 minutes to steam the rice.

4. **Finish the dish**. Unlock and open the cooker. Stir before serving.

Wild *and* Brown Rice Pilaf

PREP: 5 MINUTES • PRESSURE: 27 MINUTES • TOTAL: 32 MINUTES • PRESSURE LEVEL: HIGH • RELEASE: COMBINATION

SERVES 4

Ingredients

1 tablespoon olive oil

¾ cup diced onion

1 garlic clove, minced

⅓ cup wild rice

⅔ cup water

½ teaspoon kosher salt, divided, plus additional for seasoning

½ cup brown rice

¾ cup low-sodium vegetable broth

¼ cup dry white wine

1 bay leaf

1 fresh thyme sprig, or ¼ teaspoon dried thyme

2 tablespoons chopped fresh parsley

Directions

1. **Preparing the Ingredients**. Set the Crock-Pot Express® to "brown," heat the olive oil until it shimmers and flows like water. Add the onion and garlic, and cook for about 3 minutes, stirring, until the garlic is fragrant and the onions soften and separate. Add the wild rice, water, and ¼ teaspoon of kosher salt, and stir.

5. **High pressure for 15 minutes**. Lock the lid in place, and bring the pot to high pressure. Cook at high pressure for 15 minutes. To get 15-minutes cook time, press "Rice/Risotto" button and use the TIME ADJUSTMENT button to adjust the cook time to 15 minutes.

2. **Pressure Release.** Use the quick-release method.

3. Unlock and remove the lid. Stir in the brown rice, vegetable broth, remaining ¼ teaspoon of kosher salt, white wine, bay leaf, and thyme.

6. **High pressure for 12 minutes**. Lock the lid in place, and bring the pot back to high pressure. Cook at high pressure for 12 minutes. To get 12-minutes cook time, press "Rice/Risotto" button. When the timer goes off, turn the cooker off. ("warm" setting, turn off).

4. **Pressure Release**. After cooking, use the **natural method** to release pressure for 12 minutes, then the **quick method** to release the remaining pressure.

5. **Finish the dish**. Unlock and remove the lid. Remove the bay leaf and thyme sprig, and stir in the parsley. Taste and adjust the seasoning, as needed. Replace but *do not lock* the lid. Let the rice steam for about 4 minutes, fluff gently with a fork, and serve.

PER SERVING: CALORIES: 195; FAT: 4G; SODIUM: 309MG; CARBOHYDRATES: 32G; FIBER: 2G; PROTEIN: 5G

Spanish Rice

PREP: 5 MINUTES • PRESSURE: 8 MINUTES • TOTAL: 13 MINUTES • PRESSURE LEVEL: HIGH • RELEASE:QUICK
SERVES 4

Ingredients
- 1 tablespoon olive oil
- ⅓ cup chopped onion
- 1 large garlic clove, minced
- 1 small jalapeño pepper, seeded and chopped (about 1 tablespoon)
- ½ teaspoon kosher salt, plus additional for seasoning
- ¾ cup long-grain white rice
- ¼ cup plus 2 tablespoons Red Table Salsa
- ¾ cup low-sodium vegetable broth
- 1 tablespoon chopped fresh parsley

Directions

1. **Preparing the Ingredients** Turn the Crock-Pot Express® to "brown," heat the olive oil until it shimmers and flows like water. Add the onion, garlic, and jalapeño, and sprinkle with a pinch or two of kosher salt. Cook for about 3 minutes, stirring, until the vegetables just begin to brown. Add the rice, and stir to coat with the olive oil. Add the Red Table Salsa, and cook for 30 seconds, stirring. Add the vegetable broth and ½ teaspoon of kosher salt, and stir to combine.

7. **High pressure for 8 minutes** Lock the lid in place, and bring the pot to high pressure. Cook at high pressure for 8 minutes. To get 8-minutes cook time, press "Rice/Risotto" button and use the TIME ADJUSTMENT button to adjust the cook time to 8 minutes.

2. **Pressure Release**. Use the quick-release method.

3. **Finish the dish.** Unlock and remove the lid. Gently stir in the parsley, and replace but *do not lock* the lid. Let the rice sit for 4 to 5 minutes, fluff with a fork, and serve.

4. **Enjoy!**

PER SERVING: CALORIES: 174; FAT: 4G; SODIUM: 451MG; CARBOHYDRATES: 31G; FIBER: 1G; PROTEIN: 4G

Wild Rice Salad with Apples

PREP: 5 MINUTES • PRESSURE: 18 MINUTES • TOTAL: 23 MINUTES • PRESSURE LEVEL: HIGH • RELEASE: NATURAL

SERVES 4

Ingredients

4 cups water

1¼ teaspoons kosher salt, divided

1 cup wild rice

⅓ cup walnut or olive oil

3 tablespoons cider vinegar

¼ teaspoon celery seed

⅛ teaspoon freshly ground black pepper

Pinch granulated sugar

½ cup walnut pieces, toasted

2 or 3 celery stalks, thinly sliced (about 1 cup)

1 medium Gala, Fuji, or Braeburn apple, cored and cut into ½-inch pieces

Directions

1. **Preparing the Ingredients**. Add the water into the Crock-Pot Express®, and 1 teaspoon of kosher salt. Stir in the wild rice.
8. **High pressure for 18 minutes**. Lock the lid in place, and bring the pot to high pressure. Cook at high pressure for 18 minutes. To get 18-minutes cook time, press "Rice/Risotto" button and use the TIME ADJUSTMENT button to adjust the cook time to 18 minutes.
2. **Pressure Release**. Use the natural-release method.
3. **Finish the dish.** Unlock and remove the lid. The rice grains should be mostly split open. If not, simmer the rice for several minutes more, in the Crock-Pot Express® set to "Sauté/brown" until at least half the grains have split. Drain and cool slightly.

   To a small jar with a tight-fitting lid, add the walnut oil, cider vinegar, celery seed, the remaining ¼ teaspoon of kosher salt, the pepper, and the sugar, and shake until well combined.

   To a medium bowl, add the cooled rice, walnuts, celery, and apple. Pour half of the dressing over the salad, and toss gently to coat, adding more dressing as desired. Serve.

PER SERVING: CALORIES: 335; FAT: 17G; SODIUM: 162MG; CARBOHYDRATES: 37G; FIBER: 5G; PROTEIN: 13G

Seafood Risotto

PREP: 10 MINUTES • PRESSURE: 6 MINUTES • TOTAL: 16 MINUTES • PRESSURE LEVEL: HIGH • RELEASE: NATURAL

SERVES 4

Ingredients

3 cups mixed seafood (shrimp, calamari, clams, etc.)
Water, as needed
2 tablespoons olive oil, plus more to finish
3 garlic cloves, chopped
3 oil-packed anchovies
2 cups Arborio or Carnaroli rice
Freshly squeezed juice of 1 lemon
2 teaspoons salt
¼ teaspoon ground white pepper
1 bunch flat-leaf parsley, chopped
Lemon wedges, for serving

Directions

1. **Preparing the Ingredients**. Separate the shellfish from the other seafood and set the shellfish aside. Add the remaining seafood to a 4-cup measuring cup and add water to just over the 4-cup mark. Heat the Crock-Pot Express® using the "Brown/Sauté" mode, add the oil, and heat briefly. Stir in the garlic and anchovies and sauté until the garlic is golden and the anchovies are broken up. Add the rice, stirring to coat well. While you continue to stir, look carefully at the rice, it will first become wet and look slightly transparent and pearly; then it will slowly begin to look dry and solid white again. At that point pour in the lemon juice. Scrape the bottom of the Crock-Pot Express® gently, and keep stirring until all of the juice has evaporated. Stir in the seafood and water and the salt and pepper. Place the shellfish on top without stirring any further.

2. **High pressure for 6 minutes**. Close and lock the lid of the Crock-Pot Express®. Cook at high pressure for 6 minutes. To get 6-minutes cook time, press "Rice/Risotto" button and use the TIME ADJUSTMENT button to adjust the cook time to 6 minutes.

3. **Pressure Release**. When the time is up, open the Crock-Pot Express® with the Normal Release method .

4. **Finish the dish**. Stir the risotto. Swirl some oil over the top and sprinkle with parsley. Serve with lemon wedges.

Brown Rice Pilaf With Cashews

PREP: 5 MINUTES • PRESSURE: 33 MINUTES • TOTAL: 38 MINUTES • PRESSURE LEVEL: HIGH • RELEASE: QUICK

SERVES 6

Ingredients

3 tablespoons unsalted butter

1 large leek, white and pale green parts only, halved lengthwise, washed, and thinly sliced

½ teaspoon dried thyme

½ teaspoon salt

⅛ teaspoon ground turmeric

1½ cups long-grain brown rice, such as brown basmati

3 cups vegetable or chicken broth

½ cup chopped roasted unsalted cashews

Directions

1. **Preparing the Ingredients**
2. **High pressure for 50 minutes**.
3. **Pressure Release**
4. **Finish the dish.**

1. **Preparing the Ingredients**. Melt the butter in the Crock-Pot Express® turned to the "Brown/Sauté" function. Add the leek and cook, stirring often, until softened, about 2 minutes. Stir in the thyme, salt, and turmeric until fragrant, less than half a minute. Add the rice and cook for 1 minute, stirring all the while. Pour in the broth and stir well to get any browned bits off the bottom of the cooker.

2. **High pressure for 33 minutes**. Lock the lid onto the pot. Set the Crock-Pot Express® to cook at high pressure for 33 minutes. To get 33-minutes cook time, press "Rice/Risotto" button and use the TIME ADJUSTMENT button to adjust the cook time to 33 minutes.

3. **Pressure Release**. Use the quick-release method to return the pot's pressure to normal but do not open the cooker. Set aside for 10 minutes to steam the rice.

4. **Finish the dish**. Unlock and open the pot. Stir in the chopped cashews before serving.

Indian Veggie Pullow

PREP: 11 MINUTES • PRESSURE: 3 MINUTES • TOTAL: 14 MINUTES • PRESSURE LEVEL: HIGH • RELEASE: NATURAL
SERVES 4

Ingredients

- ½ cup cashews
- 2 cups basmati rice
- 4 tablespoons ghee or vegetable oil
- 1 large onion, finely chopped
- 3 cardamom pods, lightly crushed
- 3 or 4 whole cloves, lightly crushed
- 1 tablespoon smashed garlic
- 1 tablespoon peeled and grated fresh ginger
- ½ teaspoon crushed red pepper flakes
- 1 teaspoon ground coriander
- ½ teaspoon ground turmeric
- ½ teaspoon ground cinnamon
- 1 cup frozen petite green peas
- 1 cup coarsely chopped cauliflower florets (1-inch pieces)
- 2 carrots, peeled and diced
- 3 cups water
- 2 teaspoons salt

Directions

1. **Preparing the Ingredients.** Toast the cashews in a dry skillet over low heat until golden. Place the rice in a fine-mesh strainer and rinse it. Rest the strainer with the rice in a bowl and cover with water to soak. Meanwhile, heat the Crock-Pot Express® using the "Brown/Sauté" Mode, add the ghee, and heat briefly. Stir in the onion and fry until golden, about 7 minutes. Stir in the cardamom and cloves and sauté for about 1 minute. Lift the strainer so the rice can drain. Add the garlic, ginger, red pepper flakes, coriander, turmeric, and cinnamon to the onion mixture and sauté for another 30 seconds. Then add the peas, cauliflower, carrots, and rice; mix well. Sauté for about 3 more minutes. Stir in the water and salt.
5. **High pressure for 3 minutes**. Close and lock the lid of the Crock-Pot Express®. Cook at high pressure for 3 minutes. To get 3-minutes cook time, press "Steam" button and use the TIME ADJUSTMENT button to adjust the cook time to 3 minutes.
2. **Pressure Release**. When the time is up, open the Crock-Pot Express® with the 10-Minute Natural Release method.
3. **Finish the dish**. Fluff the pullow with a fork; taste and add more salt if you wish. Sprinkle the cashews over the top.

Brown Rice with Lentils

PREP: 5 MINUTES • PRESSURE: 35 MINUTES • TOTAL: 40 MINUTES • PRESSURE LEVEL: HIGH • RELEASE: NATURAL

SERVES 8

Ingredients

- 5 tablespoons olive oil
- 3 large onions, halved through the root (flatter) end, then sliced into thin half-moons
- 1 teaspoon coriander seeds
- 1 teaspoon cumin seeds
- ½ teaspoon ground turmeric
- ½ teaspoon ground allspice
- ½ teaspoon ground cinnamon
- 2 cups long-grain brown rice, preferably basmati
- 1 teaspoon sugar
- 1 teaspoon ground black pepper
- ½ teaspoon salt
- 4½ cups vegetable or chicken broth
- ½ cup green lentils (French lentils or lentils de Puy)

Directions

1. **Preparing the Ingredients.** Heat 1½ tablespoons oil in the Crock-Pot Multi-Cooker® turned to the "Browning" function. Add half the onions and cook until well browned and crisp at the edges, at least 10 minutes, stirring occasionally. Transfer the cooked onions to a large bowl; repeat with 1½ tablespoons more oil and the rest of the onions. Add the remaining 2 tablespoons oil to the cooker; stir in the coriander, cumin, turmeric, allspice, and cinnamon until aromatic, about 1 minute. Add the rice, sugar, pepper, and salt; stir for 1 minute. Stir in the broth, scraping up any brown bits in the cooker. Stir in the lentils.

2. **High pressure for 35 minutes.** Lock the lid on the Crock-Pot Multi-Cooker® and then cook for 35 minutes. To get 35-minutes cook time, press "Soup button and use the TIME ADJUSTMENT button to adjust the cook time to 35 minutes.

3. **Pressure Release.** Turn off the Crock-Pot Multi-Cooker® or unplug it so it doesn't flip to the keep-warm setting. Let its pressure return normal naturally, 14 to 20 minutes.

4. **Finish the dish.** Unlock and open the cooker. Spoon the caramelized onions on top of the rice; set the lid back on the cooker without locking it in place, and set aside for 10 minutes to warm the onions. Serve by scooping up big spoonfuls with onions and rice in each.

Tiger Prawn Risotto

PREP: 10 MINUTES • PRESSURE: 30 MINUTES • TOTAL: 40 MINUTES • PRESSURE LEVEL: HIGH • RELEASE: NATURAL

SERVES 2-4

Ingredients

- ½ pound frozen tiger prawns, thawed and peeled
- 1 teaspoon salt
- 1 teaspoon white pepper
- 3 tablespoons olive oil
- 4 tablespoons butter
- 1 shallot, minced
- 3 cloves garlic, minced
- 2 cups Arborio rice
- ¾ cup cooking sake
- 2 teaspoons soy sauce
- 4 cups fish stock or Japanese Dashi
- 20 grams Parmesan cheese, finely grated
- 2 stalk green onions, thinly sliced

Directions

1. **Preparing the Ingredients.** In mixing bowl season the prawns with salt and white pepper. Set the Crock-Pot Multi-Cooker® on brown and add the olive oil and butter and sauté prawns for 5-10 minutes with the shallot and garlic, the prawns should be about 80% cooked. Remove and set aside.

   Add the Arborio rice, cooking sake, soy sauce and fish stock into Crock-Pot Multi-Cooker® with a swirl of olive oil. Stir and combine, make sure the rice is coated with the liquids or Japanese Dashi

2. **High pressure for 25 minutes.** Lock the lid on the Crock-Pot Multi-Cooker® and then cook for 25 minutes. To get 25-minutes cook time, press Rice/Risotto button and use the TIME ADJUSTMENT button to adjust the cook time to 25 minutes.

3. **Pressure Release** Use the quick-release method to return the pot's pressure to normal. Place the prawns on top of the risotto and sprinkle the Parmesan cheese over the prawns and risotto.

4. **High pressure for 5 minutes.** Cover and lock the lid again and cook on High for another 5 minutes. To get 5-minutes cook time, press Beans/Chili button and use the TIME ADJUSTMENT button to adjust the cook time to 5 minutes.

5. **Pressure Release** Use the quick-release method to return the pot's pressure to normal.

6. **Finish the dish.** Garnish with the sliced green onions. Serve and Enjoy!

Main Dishes – Beans And Grains

Basic Boiled Beans

PREP: 5 MINUTES • PRESSURE: 12 MINUTES • TOTAL: 17 MINUTES • PRESSURE LEVEL: HIGH • RELEASE: QUICK

SERVES 5

Ingredients

2 cups dried beans, soaked, rinsed, and drained
4 cups water
1 herb, choose one: bay leaf, fresh thyme sprig, fresh sage sprig
1 to 3 aromatics, choose one from each: garlic, onion, shallots carrot, bell pepper celery, fennel, green bell pepper, parsley stems
1 tablespoon fat, choose one: vegetable oil, butter, margarine, rendered fat, 1 thick slice of bacon

Directions

1. **Preparing the Ingredients.** Place the beans in the Crock-Pot Express® base and add the water. Then add your chosen herb, aromatics, and fat.

2. **High pressure for 5-15 minutes**. Close and lock the lid of the Crock-Pot Express®. Cook at high pressure for the time appropriate for the type of bean and cooker, 5 to 15 minutes.
3. **Pressure Release** When the time is up, open the Crock-Pot Express® with the Natural Release method; this should take 20 to 30 minutes.
4. **Finish the dish.** Drain the beans in a colander set over a bowl. Remove the herb and aromatics and reserve the cooking liquid to use in place of stock at a later time. Use the beans in any recipe calling for cooked beans.

One Minute Quinoa

PREP: 5 MINUTES • PRESSURE: 1 MINUTES • TOTAL: 6 MINUTES • PRESSURE LEVEL: HIGH • RELEASE: NATURAL
SERVES 6-8

Ingredients

2 cups whole-grain quinoa, rinsed and drained
3 cups water
2 teaspoons salt

Directions

1. **Preparing the Ingredients.** Place the quinoa, water, and salt in the Crock-Pot Express® base.
2. **High pressure for 1 minute**. Close and lock the lid of the Crock-Pot Express®. Cook at high pressure for 1 minute. Press "Steam"Button, press "Cancel" button after 1 Minute.
3. **Pressure Release.** Open the Crock-Pot Express® with the 10-Minute Natural Release method.
4. **Finish the dish.** Fluff the quinoa with a fork and serve.

Franks And Beans

PREP: 7 MINUTES • PRESSURE: 7 MINUTES • TOTAL: 14 MINUTES • PRESSURE LEVEL: HIGH • RELEASE: QUICK
SERVES 4

Ingredients

3 ounces slab bacon, chopped

1 medium yellow onion, chopped

2 teaspoons minced garlic

Two 15-ounce cans pinto beans, drained and rinsed (about 3½ cups)

½ cup ketchup

¼ cup maple syrup, preferably Grade B or 2

2 tablespoons Dijon mustard

2 tablespoons packed dark brown sugar

½ teaspoon ground black pepper

¼ teaspoon ground cloves

1 pound hot dogs, cut into 2-inch pieces

Directions

1. **Preparing the Ingredients.** Fry the bacon until brown and crisp, stirring occasionally, in the Crock-Pot Express® turned to the "Browning" function, about 3 minutes.
   Add the onion and cook until softened, about 4 minutes, stirring occasionally. Add the garlic and cook until aromatic, less than 1 minute. Stir in the beans, ketchup, maple syrup, mustard, brown sugar, pepper, and cloves until the brown sugar dissolves; then stir in the hot dogs.
2. **High pressure for 7 minutes.** Lock the lid onto the pot. Set the Crock-Pot Express® to cook at high pressure for 7 minutes. To get 7 minutes cook time, press the "Steam" button and use the TIME ADJUSTMENT button to adjust the cook time to 7 minutes.
3. **Pressure Release** Use the quick-release method.
4. **Finish the dish** Unlock and open the pot. Stir well before serving.

Pinto Beans With Bacon

PREP: 5 MINUTES • PRESSURE: 18 MINUTES • TOTAL: 23 MINUTES • PRESSURE LEVEL: HIGH • RELEASE: QUICK
SERVES 4

Ingredients
1 cup dried pinto beans
1 tablespoon unsalted butter
3 thin bacon slices, chopped
½ cup chopped pecans
1 medium yellow or white onion, chopped
½ teaspoon dried oregano
½ teaspoon ground cumin
¼ teaspoon ground coriander
One 4½-ounce can chopped mild green chiles (about ½ cup)

Directions

1. **Preparing the Ingredients** Soak the beans in a large bowl of water on the counter overnight, for at least 12 hours or up to 16 hours.
   Drain the beans in a colander set in the sink; pour them into a 6-quart Crock-Pot Express®. Add enough cool tap water that they're submerged by 2 inches.
2. **High pressure for 18 minutes.** Lock the lid onto the pot. Set the Crock-Pot Express® to cook at high pressure for 18 minutes. To get 18 minutes cook time, press the "Beans/Chili" button and use the TIME ADJUSTMENT button to adjust the cook time to 18 minutes
3. **Pressure Release** Use the quick-release method.
4. **Finish the dish.** Unlock and open the cooker. Scoop out 1 cup of the cooking liquid and reserve it. Drain the beans in a colander set in the sink. Wipe out the cooker.
5. Melt the butter in the Crock-Pot Express® turned to its "browning" function. Add the bacon and pecans; fry until both are lightly browned, stirring occasionally, about 3 minutes. Add the onion and cook, stirring often, until softened, about 3 minutes.
6. Stir in the oregano, cumin, and coriander until aromatic, about 20 seconds. Then pour in the drained beans, green chiles, and ¼ cup of the reserved cooking liquid. Cook, stirring often, until the beans are just heated through, adding more of the reserved cooking liquid in ¼-cup increments when the mixture gets too dry.
7. Serve and Enjoy

Spicy Black-Eyed Peas

PREP: 5 MINUTES • PRESSURE: 22 MINUTES • TOTAL: 27 MINUTES • PRESSURE LEVEL: HIGH • RELEASE: QUICK

SERVES 6

Ingredients

4 ounces slab bacon, chopped

1 medium yellow onion, chopped

2 cups dried black-eyed peas

One 14-ounce can diced tomatoes (about 1¾ cups)

One 4½-ounce can chopped hot green chiles (about ½ cup)

1 tablespoon dried oregano

Directions

1. **Preparing the Ingredients** Fry the bacon in the Crock-Pot Express® turned to its "browning" function, until it begins to brown and give off its fat, about 2 minutes. Add the onion and cook, stirring often, until it turns translucent, about 4 minutes. Pour in 3 cups water; add the black-eyed peas, tomatoes, chiles, and oregano, and stir well.
2. **High pressure for 22 minutes.** Lock the lid onto the cooker. Set the Crock-Pot Express® to cook at high for 22 minutes. To get 22 minutes cook time, press the "Beans/Chili" button and use the TIME ADJUSTMENT button to adjust the cook time to 22 minutes
3. **Pressure Release** Use the quick-release method.
4. **Finish the dish.** Unlock and open the cooker. Stir well before serving.
5. Serve and Enjoy!

Frijoles Refritos

PREP: 5 MINUTES • PRESSURE: 50 MINUTES • TOTAL: 55 MINUTES • PRESSURE LEVEL: HIGH • RELEASE: NATURAL
SERVES 4

Ingredients

4 cups water
1 teaspoon kosher salt
2 bacon slices, each cut into 2 or 3 pieces, divided
1 very small onion, peeled, trimmed, and halved through the root
½ pound dried pinto beans, rinsed and picked over
2 garlic cloves, smashed

Directions

1. **Preparing the Ingredients** Add the water into the Crock-Pot Express®, 1 teaspoon kosher salt, 1 bacon slice, and the onion. Pour in the beans.
   Lock the lid in place, and bring the pot to high pressure.
2. **High pressure for 50 minutes** Cook at high pressure for 50 minutes. To get 50 minutes cook time, press the "Multigrain" button and use the TIME ADJUSTMENT button to adjust the cook time to 50 minutes

3. **Pressure Release** Use the natural method to release pressure.
4. **Finish the dish.** Unlock and remove the lid. Working over a large bowl, place a large sieve or colander, and drain the beans. The onion and bacon pieces will likely have dissolved into the beans, but if there are any large pieces left, remove and discard them. Reserve the liquid.
   Turn the Crock-Pot Express® to "brown." Add the remaining bacon slice, and cook until the pieces are crisp and the fat renders. Remove the bacon, and set aside, leaving the fat in the cooker. Add the garlic to the fat, and cook for about 6 minutes, or until well browned or a bit charred. Remove the garlic and discard.
   Pour the beans back into the cooker. Using a large spoon or potato masher, mash the beans into the bacon fat. Stir the reserved bean liquid to redistribute the starch, and add ¼ cup of the liquid to the beans. Continue mashing the beans until they're a rough purée, or the texture you prefer. Add additional bean liquid, as necessary, to keep the beans from drying out. Chop the reserved bacon, sprinkle over the top of the beans, and serve.

PER SERVING: CALORIES: 261; FAT: 5G; SODIUM: 525MG; CARBOHYDRATES: 39G; FIBER: 9G; PROTEIN: 16G

"Baked" Beans

PREP: 5 MINUTES • PRESSURE: 10 MINUTES • TOTAL: 15 MINUTES • PRESSURE LEVEL: HIGH • RELEASE: QUICK
SERVES 4

Ingredients

1 tablespoon kosher salt
1 quart water
½ pound dried navy beans
1 teaspoon olive oil
1 bacon or ham slice, diced
½ cup chopped onion
⅓ cup diced green bell pepper
1¾ cups Chicken Stock or low-sodium broth
¼ cup ketchup
¼ cup packed brown sugar
2 teaspoons cider vinegar
2 teaspoons Worcestershire sauce
¼ teaspoon dried mustard

Directions

1. **Preparing the Ingredients.** In a large bowl, dissolve 1 tablespoon of kosher salt in the water. Add the beans, and soak at room temperature for 8 to 24 hours. Drain and rinse.
   Turn the Crock-Pot Express® to "brown," add the olive oil and bacon. Cook for about 3 minutes, stirring, until the bacon begins to crisp. Add the onion and green bell pepper, and cook for 2 to 3 minutes, until softened. Add the Chicken Stock to the Crock-Pot Express®; then pour in the beans.
   Lock the lid in place, and bring the pot to high pressure.

2. **High pressure for 10 minutes**. Cook at high pressure for 10 minutes. To get 10 minutes cook time, press the "Steam" button.

3. **Pressure Release** Use the quick-release method.

4. **Finish the dish.** Unlock and remove the lid. The beans should barely be done—cooked but quite firm. Add the ketchup, brown sugar, vinegar, Worcestershire sauce, and dried mustard, and stir to combine. Turn the Crock-Pot Express® to "simmer." Cook for about 15 minutes, uncovered, or until the sauce has thickened and the beans are completely tender, and serve.

5. For real baked beans, pour the beans with their sauce into a baking dish at step 5. Bake at 350°F for 25 to 40 minutes, or until the sauce has thickened and the beans are tender.

PER SERVING: CALORIES: 281; FAT: 3G; SODIUM: 614MG; CARBOHYDRATES: 50G; FIBER: 15G; PROTEIN: 15G

White Beans with Prosciutto

PREP: 5 MINUTES • PRESSURE: 12 MINUTES • TOTAL: 17 MINUTES • PRESSURE LEVEL: HIGH • RELEASE: QUICK
SERVES 4

Ingredients
1 tablespoon kosher salt
1 quart water
½ pound dried navy beans
1 tablespoon olive oil
3 ounces prosciutto or ham, diced
2 medium garlic cloves, minced
1¾ cups Chicken Stock or low-sodium broth
1 or 2 fresh rosemary sprigs
2 tablespoons dry white wine

Directions

1. **Preparing the Ingredients** In a large bowl, dissolve the kosher salt in the water. Add the beans, and soak at room temperature for 8 to 24 hours. Drain and rinse. Set the Crock-Pot Express® to its "brown" function, heat the olive oil until it shimmers and flows like water. Add the prosciutto, and cook for about 3 minutes, stirring, until the prosciutto starts to crisp. Add the garlic, and cook for 1 more minute, or until fragrant. Add the Chicken Stock and rosemary to the Crock-Pot Express®, then pour in the beans.
Lock the lid in place, and bring the pot to high pressure.

2. **High pressure for 12 minutes**. Cook at high pressure for 12 minutes. To get 12 minutes cook time, press the "Rice/Risotto" button.
3. **Pressure Release** Use the quick-release method.
4. **Finish the dish** Unlock and remove the lid. Remove the rosemary sprig, but don't worry if some of the needles have fallen off; they'll be tender enough to eat. If the beans are too soupy, turn the Crock-Pot Express® to "simmer," and simmer until some of the liquid has evaporated. Stir in the white wine, and serve.

PER SERVING: CALORIES: 271; FAT: 6G; SODIUM: 602MG; CARBOHYDRATES: 36G; FIBER: 14G; PROTEIN: 17G

Chocolate Pudding

PREP: 5 MINUTES • PRESSURE: 15 MINUTES • TOTAL: 20 MINUTES • PRESSURE LEVEL: HIGH • RELEASE: NATURAL

SERVES 6

Ingredients

6 ounces semisweet or bittersweet chocolate, chopped

½ ounce unsweetened chocolate, chopped

6 tablespoons sugar

1½ cups light cream

4 large egg yolks, at room temperature and whisked in a small bowl

1 tablespoon vanilla extract

¼ teaspoon salt

Directions

1. **Preparing the Ingredients.** Place all the chopped chocolate and the sugar in a large bowl. Heat the cream in a saucepan over low heat until small bubbles fizz around the inside edge of the pan.

   Pour the warmed cream over the chocolate; whisk until the chocolate has completely melted. Cool a minute or two, then whisk in the yolks, vanilla, and salt. Pour the mixture into six ½-cup heat-safe ramekins, filling each about three-quarters full. Cover each with foil.

   Set the rack in the Crock-Pot Express®; pour in 2 cups water. Set the ramekins on the rack, stacking them as necessary without any one ramekin sitting directly on top of another.

   Lock the lid onto the pot.

2. **High pressure for 15 minutes**. Set the Crock-Pot Express® to cook at high pressure for 15 minutes. To get 15 minutes cook time, press the "Poultry" button.

3. **Pressure Release** Turn off the Crock-Pot Express® or unplug it so it doesn't flip to its keep-warm setting. Let its pressure return to normal naturally, 10 to 14 minutes.

4. **Finish the dish.** Unlock and open the cooker. Transfer the hot ramekins to a cooling rack, uncover, and cool for 10 minutes before serving—or chill in the refrigerator for up to 3 days, covering again once the puddings have chilled.

Chocolate Brownie

PREP: 5 MINUTES • PRESSURE: 15 MINUTES • TOTAL: 20 MINUTES • PRESSURE LEVEL: HIGH • RELEASE: QUICK
SERVES 2

Ingredients

2 tablespoons unsalted butter
1 tablespoon dark chocolate chips
⅓ cup granulated sugar
1 egg
⅛ teaspoon vanilla extract
¼ cup all-purpose flour
2 tablespoons cocoa powder
1 cup water, for steaming (double-check the Crock-Pot Express® manual to confirm amount, and follow the manual if there is a discrepancy)
1 tablespoon confectioners' sugar or powdered sugar

Directions

1. **Preparing the Ingredients.** In a small microwave-safe bowl, microwave the butter and chocolate chips for 30 seconds on high to melt. Into a small mixing bowl, scrape the chocolate mixture, and add the sugar. Beat for about 2 minutes. Add the egg and vanilla, and beat for about 1 minute more, until smooth. Sift the flour and cocoa powder over the wet ingredients, and beat until just combined.
   Spoon the batter into a nonstick mini springform pan (4½ inches) or a mini loaf pan (3-by-5-inch), and smooth the top.
   Add the water into the Crock-Pot Express®, and insert the steamer basket or trivet. Place the loaf pan on the steamer insert. Place a square of aluminum foil over the pan, but don't crimp it down; it's just to keep steam from condensing on the surface of the cake.
2. **High pressure for 15 minutes**. Lock the lid in place, and bring the pot to high pressure. Cook at high pressure for 15 minutes. To get 15 minutes cook time, press the "Poultry" button.
3. **Pressure Release** Use the quick-release method.
4. **Finish the dish.** Unlock and remove the lid. Using tongs, remove the sheet of foil. Transfer the pan to a cutting board or rack to cool. Dust the cake with the confectioners' sugar, slice, and serve.

PER SERVING: CALORIES: 370; FAT: 16G; SODIUM: 114MG; CARBOHYDRATES: 58G; FIBER: 2G; PROTEIN: 6G

Chai-Spiced Apricot Crisp

PREP: 5 MINUTES • PRESSURE: 15 MINUTES • TOTAL: 20 MINUTES • PRESSURE LEVEL: HIGH • RELEASE: NATURAL

SERVES 6

Ingredients

  5 tablespoons cold unsalted butter
  2 pounds fresh apricots, pitted and diced
  ½ cup finely ground bread crumbs
  ½ cup turbinado (raw) sugar
  ½ cup plain whole-milk yogurt
  1 tablespoon chopped crystallized ginger
  ⅛ teaspoon ground cardamom
  ¼ teaspoon ground cinnamon
  Pinch ground white pepper
  6 tablespoons granulated sugar
  Vanilla ice cream, for serving

Directions

1. **Preparing the Ingredients.** Add 2 cups of water to the Crock-Pot Express® base; insert the steamer basket and set aside.

   Chop 3 tablespoons of the butter and place in a 4-cup heat-proof bowl; add the apricots, half the bread crumbs, the turbinado sugar, yogurt, ginger, cardamom, cinnamon, and pepper; mix well. Using a foil sling, lower the bowl into the Crock-Pot Express®; do not cover the bowl.

   Close and lock the lid of the Crock-Pot Express®.

2. **High pressure for 15 minutes** Cook at high pressure for 15 minutes. To get 15 minutes cook time, press the "Poultry" button.

3. **Pressure Release** Use the Normal Release method.

   Meanwhile, chop the remaining 2 tablespoons butter. Turn on the broiler.

4. **Finish the dish.** When the cooker is open, lift the heat-proof bowl out of the cooker. Moving quickly, sprinkle the granulated sugar and remaining ¼ cup bread crumbs over the apricot mixture, then scatter the chopped butter over the top.

5. Place the heat-proof bowl under the broiler until the top of the crisp is caramelized and turns golden brown, about 5 minutes. Serve warm in small bowls with a scoop of vanilla ice cream.

   Cakes

   The Crock-Pot Express® gives an unexpected bonus to cake baking. The process is much faster than usual because there is no need to preheat the oven and the results are absolutely moist!

Pumpkin Pie Pudding

PREP: 5 MINUTES • PRESSURE: 22 MINUTES • TOTAL: 27 MINUTES • PRESSURE LEVEL: HIGH • RELEASE: NATURAL

SERVES 6

Ingredients

1½ cups canned pumpkin

½ cup packed dark brown sugar

½ cup heavy cream

2 large eggs, at room temperature

2 tablespoons unsulphured molasses

1 teaspoon vanilla extract

2 tablespoons all-purpose flour

1 teaspoon ground cinnamon

¼ teaspoon salt

Directions

1. **Preparing the Ingredients.** Lightly butter the inside of a 1-quart round, high-sided soufflé or baking dish; set aside.

   Whisk the pumpkin, brown sugar, cream, eggs, molasses, and vanilla in a large bowl until the brown sugar has dissolved. Whisk in the flour, cinnamon, and salt until smooth.

   Pour the mixture into the prepared baking dish. Butter one side of a 10-inch piece of aluminum foil and set it buttered side down over the baking dish; seal well.

   Set the Crock-Pot Express® rack; pour in 2 cups water. Crimp the ends of the sling to fit inside the cooker.

   Lock the lid onto the pot.

2. **High pressure for 22 minutes**. Set the Crock-Pot Express® to cook at high for 22 minutes. To get 22 minutes cook time, press the "Beans/Chili" button and use the TIME ADJUSTMENT button to adjust the cook time to 22 minutes

3. **Pressure Release** Turn off the Crock-Pot Express® or unplug it so it doesn't flip to its keep-warm setting. Let its pressure return to normal naturally, 10 to 14 minutes.

4. **Finish the dish.** Unlock and open the cooker. Use the foil sling to transfer the baking dish to a wire cooling rack; uncover and set aside for a few minutes, until the pudding is firm and set.

5. Serve by dishing it up warm by the spoonful.

Amaretti-Stuffed Apples

PREP: 5 MINUTES • PRESSURE: 5 MINUTES • TOTAL: 10 MINUTES • PRESSURE LEVEL: HIGH • RELEASE: NATURAL

SERVES 4-6

Ingredients

- 4 to 6 very small apples, preferably Granny Smith
- 4 to 6 tablespoons lemon marmalade, or your favorite preserves
- 1¼ cups crumbled amaretti cookies
- ½ cup crushed walnuts
- 3 teaspoons turbinado (raw) sugar
- 4 tablespoons cold unsalted butter, chopped
- Vanilla ice cream, for serving

Directions

1. **Preparing the Ingredients** Add 2 cups of water to the Crock-Pot Express® base and set aside.

   Slice the apples in half lengthwise and then use a melon bailer to scoop out the core from each half. Leave the stem on—it looks pretty. Paint the cut surface of each with a little of the marmalade to keep it from oxidizing.

   Add 1 cup of the cookie crumbs, the walnuts, and sugar to a small bowl and mix in the butter until well combined. Spoon the mixture onto each apple half, filling the cavity and covering the top. Arrange the apples in the steamer basket. If you run out of room, you can either make a second row, offset on the first pyramid-style, or hold the extras to cook as a second batch. Lower the steamer basket into the Crock-Pot Express®.

2. **High pressure for 5 minutes** Close and lock the lid of the Crock-Pot Express® . Cook at high pressure for 5 minutes. To get 5 minutes cook time, press the "Steam" button and use the TIME ADJUSTMENT button to adjust the cook time to 5 minutes

3. **Pressure Release** When the time is up, open the Crock-Pot Express® using the Normal Release

4. **Finish the dish.** Using two spoons, delicately lift the apples from the Crock-Pot Express® and position 2 halves on each dessert plate. Serve with a dollop of vanilla ice cream and a sprinkling of the remaining cookie crumbs.

Strawberry Shortcake Mug Cake

PREP: 5 MINUTES • PRESSURE: 12 MINUTES • TOTAL: 17 MINUTES • PRESSURE LEVEL: HIGH • RELEASE: QUICK
SERVES 2

Ingredients

1 egg

½ cup (48 g) almond flour

½ tsp 100% vanilla extract

1 tbsp. (15 ml) maple syrup

1 tbsp. (15 ml) ghee

3 tbsp. (24 g) chopped strawberries (plus more for garnish)

1 cup (240 ml) water

3 tbsp. (45 ml) coconut whipped cream to garnish

Directions

1. **Preparing the Ingredients.** Combine all of the ingredients, except for the water and whipped cream, into a heat-resistant ceramic coffee mug.
Pour the cup of water into the stainless steel Crock-Pot Express® bowl and place the wire rack into the basin. Set your mug on top of the rack and secure the lid.

2. **High pressure for 12 minutes.** Seal off the pressure valve and Cook at high pressure for 12 minutes. To get 12 minutes cook time, press the "Dessert" button and use the TIME ADJUSTMENT button to adjust the cook time to 12 minutes.

3. **Pressure Release** use quick-releasing the pressure valve when the cycle is complete. Remove the lid when safe to do so and carefully remove the hot mug. Top with coconut whipped cream and additional fresh strawberries if desired.

White Chocolate Lemon Pudding

PREP: 5 MINUTES • PRESSURE: 15 MINUTES • TOTAL: 20 MINUTES • PRESSURE LEVEL: HIGH • RELEASE: NATURAL

SERVES 6

Ingredients

6 ounces white chocolate, chopped

1 cup heavy cream

1 cup half-and-half

4 large egg yolks, at room temperature and whisked in a small bowl

1 tablespoon sugar

1 tablespoon finely grated lemon zest (about 1 medium lemon)

¼ teaspoon lemon extract

Directions

1. **Preparing the Ingredients** .Put the chopped white chocolate in a large bowl. Mix the cream and half-and-half in a small saucepan and warm over low heat until bubbles fizz around the edges of the pan.

   Pour the warm mixture over the white chocolate and whisk until melted. Whisk in the egg yolks, sugar, zest, and extract. Pour the mixture into six ½-cup heat-safe ramekins; cover each tightly with aluminum foil.

   Set the Crock-Pot Express® rack in the Crock-Pot Express®; pour in 2 cups water. Set the ramekins on the rack, stacking them as necessary without any one ramekin sitting directly on top of another.

2. **High pressure for 15 minutes**. Lock the lid onto the pot. Set the Crock-Pot Express® to cook at high pressure for 15 minutes. To get 15 minutes cook time, press the "Poultry" button.

3. **Pressure Release** Turn off the Crock-Pot Express® or unplug it so it doesn't jump to its keep-warm setting. Let its pressure return to normal naturally, 10 to 14 minutes.

4. **Finish the dish** Unlock and open the cooker. Transfer the (hot!) ramekins to a cooling rack; uncover each and cool for a few minutes before serving—or store in the refrigerator for up to 3 days, covering the ramekins again after they have chilled.

Blackberry Swirl Cheesecake

PREP: 10 MINUTES • PRESSURE: 20 MINUTES • TOTAL: 20 MINUTES • PRESSURE LEVEL: HIGH • RELEASE: NATURAL

SERVES 4-6

Ingredients

1 cup fresh blackberries
½ cup powdered sugar
4 tablespoons unsalted butter
1 cup crushed graham crackers
14 ounces cream cheese (one 8-ounce and two 3-ounce packages)
½ cup granulated sugar
Freshly grated zest from 1 lemon
Freshly grated zest from half an orange
2 large eggs

Directions

1. **Preparing the Ingredients.** Add 2 cups of water to the Crock-Pot Express® base; insert the steamer basket and set aside. Cut a piece of wax paper to fit the bottom of a wide, flat-bottomed 4-cup baking dish; also cut a strip sized to fit the sides of the dish. Line the dish with the paper.

   Puree the blackberries and powdered sugar in a blender and set aside.

   Melt the butter in a medium saucepan on medium heat. Remove the pan from the heat and mix in the crushed crackers. Scoop the mixture into the prepared baking dish and, using the back of your hand, push it into a flat, thin, even layer that covers the bottom of the dish, and, if there is enough, partway up the sides. Put the dish in the refrigerator to chill, uncovered, while you prepare the filling.

   In a medium bowl, using an electric mixer on medium speed, mix together the cream cheese, granulated sugar, and lemon and orange zests. Add the eggs and mix into a smooth batter, about 5 minutes.

   Remove the dish with the crust from the refrigerator. Slowly pour the batter over the crust, spreading level. To add the blackberry swirl, pour the puree into a squirt bottle (or food storage bag with one corner clipped off) and with it draw a spiral from the center out on top of the batter. Then use a toothpick or skewer to drag radiating lines from the center to the edge of the dish. Using a foil sling, lower the dish into the Crock-Pot Express®; do not cover the dish.

2. **High pressure for 20 minutes**. Close and lock the lid of the Crock-Pot Express®. Cook at high pressure for 20 minutes. To get 20 minutes cook time, press the "Dessert" button and use the TIME ADJUSTMENT button to adjust the cook time to 20 minutes.

3. **Pressure Release.** When the time is up, open the Crock-Pot Express® using the 10-Minute Natural Release method.

4. **Finish the dish.** Lift the dish out of the Crock-Pot Express® and check the cake for doneness, transfer the dish to a wire rack.

   Let the cake cool, uncovered, for about 30 minutes. Then cover the dish with plastic wrap and refrigerate until ready to serve, for at least 4 hours.

   Work quickly and delicately to unmold the chilled cake: Invert a plate over the dish and flip the dish and plate over together. Lift the dish off the cake and then peel off the wax paper circle on the base and the strip on the sides. Then invert a serving plate on the cake and gently flip all three components over together; lift off the top plate. Serve the cake cold, cut into wedges.

Vanilla-Ginger Custard

PREP: 10 MINUTES • PRESSURE: 6 MINUTES • TOTAL: 16 MINUTES • PRESSURE LEVEL: HIGH • RELEASE: NATURAL

SERVES 2

Ingredients

⅓ cup whole milk

⅓ cup heavy (whipping) cream

½ teaspoon vanilla extract

¼ teaspoon ground ginger

2 large egg yolks

⅓ cup granulated sugar

1 cup water, for steaming (double-check the Crock-Pot Express® manual to confirm amount, and follow the manual if there is a discrepancy)

2 teaspoons chopped crystalized ginger (optional)

Directions

1. **Preparing the Ingredients** In a small saucepan set over medium heat, combine the milk, heavy cream, vanilla, and ground ginger, and bring the mixture just to a simmer. Take it off the heat and cool slightly.

   In a small bowl, whisk together the egg yolks and sugar until the sugar is dissolved and the mixture is pale yellow. Working slowly, whisk a few tablespoons of the milk mixture into the egg mixture, then repeat with a little more. Once the egg mixture is warmed, whisk in the remainder of the milk mixture.

   Pour the custard into 2 heatproof custard cups or small ramekins. Cover with aluminum foil, and crimp to seal around the edges.

   Add the water and insert the steamer basket or trivet.

   Place the custard cups on the steamer insert.

2. **High pressure for 6 minutes** Lock the lid in place, and bring the pot to high pressure. Cook at high pressure for 6 minutes. To get 6 minutes cook time, press the "Dessert" button and use the TIME ADJUSTMENT button to adjust the cook time to 6 minutes.

3. **Pressure Release** Use the natural method to release pressure.

4. **Finish the dish.** Unlock and remove the lid. Using tongs, carefully remove the custards from the cooker and remove the foil. The custards should be set but still a bit soft in the middle; they'll firm as they cool. Cool for 20 to 30 minutes, then refrigerate for several hours to chill completely. When ready to serve, top with the crystalized ginger.

5. Enjoy!

Poached Peach Cups with Ricotta and Honey

PREP: 5 MINUTES • PRESSURE: 5 MINUTES • TOTAL: 10 MINUTES • PRESSURE LEVEL: LOW • RELEASE: QUICK

SERVES 4

Ingredients

- 4 peaches, cut in half and pitted
- 1/4 cup apple juice
- 1/4 cup water
- 3 tablespoons light brown sugar
- 1/8 teaspoon ground cinnamon
- 1 cup part-skim ricotta cheese
- 2 tablespoons honey
- 1/4 teaspoon vanilla extract

Directions

1. **Preparing the Ingredients.** Add peaches, apple juice, water, brown sugar, and cinnamon to the cooker.
2. **High pressure for 5 minutes.** Securely lock the Crock-Pot Express® 's lid and set for 5minutes on LOW. To get 5 minutes cook time, press the "Dessert" button and use the TIME ADJUSTMENT button to adjust the cook time to 5 minutes.
3. **Pressure Release** Perform a quick release to release the cooker's pressure.
4. **Finish the dish.** Remove peaches from cooking liquid, and set aside.
Combine ricotta cheese, honey, and vanilla extract, and serve spooned into the center of each peach half.
Zest.

10 MOST COMMON CROCK-POT MULTI-COOKER®
MISTAKES

1. Forget to Place the Inner Pot Back into Crock-Pot Multi-Cooker® Before Pouring in Ingredients: It can be chaotic in the kitchen. We've heard many stories of users accidentally pouring ingredients into the Crock-Pot Multi-Cooker® housing without the Inner Pot.

2. Overfill the Crock-Pot Multi-Cooker®: Many new users fill their Crock-Pot Multi-Cooker® with food & liquid up to the Max Line (sometimes even a stretch over the Max Line). This may risk clogging the Venting Knob.

3. Use Quick Release For Foamy Food or When It is overfilled: Many new users are unsure when to use Quick Pressure Release and Natural Pressure Release. There's a chance of splattering if users use Quick Release when cooking foamy food, such as grains or beans.

4. Press the Timer Button to Set Cooking Time: Some new users have mistaken the "Timer" button for setting the cooking time, then wondered why the Crock-Pot Multi-Cooker® is just sitting there not doing anything.

5. Forget to Turn the Venting Knob to Sealing Position: It might be a bit overwhelming to use the Crock-Pot Multi-Cooker® in the beginning, and it's common to forget to turn the Venting Knob to the Sealing Position when cooking.

6. Place Crock-Pot Multi-Cooker® on the Stovetop and Accidentally Turned the Dial: Due to convenience or limited counter space, some users like to place their Crock-Pot Multi-Cooker® on the stovetop. Sometimes, things happen...and we see melted burnt Crock-Pot Multi-Cooker® bottom.

7. Cooking Liquid: Too Thick/Not Enough Liquid/Too Much Liquid: As a new user, it's not intuitive on how much cooking liquid to use. If there's not enough cooking liquid or the liquid is too thick, Crock-Pot Multi-Cooker® will not be able to generate enough steam to get up to pressure. On the contrary, when there is too much cooking liquid in the Crock-Pot Multi-Cooker®, it will increase the overall cooking time (both time to get up to pressure & Natural Release time). This may overcook the food.

8. Forget to Put the Sealing Ring Back in the Lid Before Cooking: Since the silicone sealing ring absorbs the food smell, many users regularly air out/wash the sealing ring. It's easy to forget to place it back into the lid before using the Crock-Pot Multi-Cooker®.

9. Use Rice Button for Cooking All Types of Rice: We've heard some new Crock-Pot Multi-Cooker® users have less than satisfactory results with their rice cooked in the Crock-Pot Multi-Cooker® using the "Rice" Button. Different types of rice require different water to rice ratios & cooking times. For best results, we like to use the "Manual" Button for most control on Cooking Method & Time.

10. Use Hot Liquid in a Recipe that Calls for Cold Liquid: Some users ran into a problem that all their Crock-Pot Multi-Cooker® meals were undercooked. We later found that they always start cooking by pouring hot water into the Crock-Pot Multi-

Cooker®. Using hot liquid in a recipe that calls for a cold liquid shortens the overall cooking time, because Crock-Pot Multi-Cooker® will take a shorter time to come up to pressure. Since the food starts to cook when Crock-Pot Multi-Cooker® is heating up & going up to pressure, and this part of the cooking time is shortened, the food may come out undercooked.

HERITAGE OF FOOD: A FAMILY GATHERING

To survive, we need to eat. As a result, food has turned into a symbol of loving, nurturing and sharing with one another. Recording, collecting, sharing and remembering the recipes that have been passed to you by your family is a great way to immortalize and honor your family. It is these traditions that carve out your individual personality. You will not just be honoring your family tradition by cooking these recipes but they will also inspire you to create your own variations, which you can then pass on to your children's.

The recipes are just passed on by everyone and nobody actually possesses them. I too love sharing recipes. The collection is vibrant and rich as a number of home cooks have offered their inputs to ensure that all of us can cook delicious meals at our home. I am thankful to each one of you who has contributed to this book and has allowed their traditions to pass on and grow with others. You guys are really wonderful!

I am also thankful to the cooks who have evaluated all these recipes. You're, as well as, the comments that came from your family members and friends were really invaluable.

If you have the time and inclination, please consider leaving a short review wherever you can, we would love to learn more about your opinion.

https://www.amazon.com/review/review-your-purchases/

About the Author

Mary is a New York-based food writer, experienced chef and passionate for life and love. She has contributed food articles in Magazines and Blogs, growing out of her commitment to make it possible for everyone to cook, even if they have too little time. Mary enjoys spending time with her husband and her kids, browsing a farmers' market, gardening and traveling.

Copyright

Made in the USA
Middletown, DE
26 January 2018